D1432972

IRAN: Economic Development under
Dualistic Conditions

Publications of the Center for Middle Eastern Studies, Number 7

WILLIAM R. POLK, GENERAL EDITOR

IRAN

Economic Development
under
Dualistic Conditions

Jahangir Amuzegar
and
M. Ali Fekrat

The University of Chicago Press
Chicago and London

Publications of the
CENTER FOR MIDDLE EASTERN STUDIES

International Standard Book Number: 0-226-01754-0
Library of Congress Catalog Card Number: 79-153044

The University of Chicago Press, Chicago 60637
The University of Chicago Press, Ltd., London

Contents

List of Tables

List of Tables

Preface

History has been Iran's everlasting claim to fame. In their many darkest hours, particularly during the last two centuries, Iranians have found inspiration and courage in their more distant and glorious past. But contemporary Iran is trying to make a history of its own. In the brief period covered in this study—brief by Iranian standards—the signs of a promising comeback have become brighter and more numerous. This study bears partial witness to a new resurgence of political self-confidence and economic progress.

The present study has two related objectives—both limited and modest in scope. One is to explore the interaction between the foreign-oriented and all-important oil sector and the other sectors of the Iranian economy in the last half century or so. The other aim is to use the Iranian experience as the basis for a more generalized development model for such dualistic economies elsewhere. This book is not an economic profile of Iran. It gives, for example, no detailed or definitive information on many aspects of the Iranian economy such as population, resources, socioeconomic institutions, specific monetary, fiscal, and commercial policies, trade relations, and income distribution. Nor is this an economic history of Iran or its oil industry. In some of these areas neglected by us there are good studies available, and more are, we hope, forthcoming.

In writing a book on the economy of Iran, one is confronted with the problem of the availability and consistency of quantitative data, the question of transliteration of Iranian names and titles, and the correspondence of the Iranian solar years with the Western Christian era. We admit that in preparing this study we have been hindered, and at times severely handicapped, by the paucity and inconsistency of available statistical data. Detailed and comparable information on the Iranian economy for the first half of the present century is almost nonexistent. This explains the rather cursory treatment of the subject in chapter 2 compared with the rest of the study. However, since 1950, and particularly in recent years, there has been great improvement in both the quantity and the quality of statistical information. Although the amount as well as the accuracy of available data for a thoroughly analytical study on Iran still leaves much to be desired, we have tried, with the aid of requisite statistical techniques, to combine many rudimentary figures scattered in official and private sources in order to reach certain limited conclusions in our analysis.

In transliterating Iranian names and titles, we have tried to reflect the present-day Persian pronunciation as closely as possible. The Iranian solar calendar, which has been the Iranian official calendar since 1925, has been converted to the Gregorian calendar by adding 621 to the solar year. Thus the Iranian year 1349 is equated with the period 21 March 1970 through 20 March 1971; and so on for other solar years.

We are indebted to many friends and scholars for their suggestions, encouragement, and assistance. Professor Ehsan Yar-Shater of Columbia University's Near East Center made us more acutely aware of the widening gap between the growing demand of American students for information on Iran's economic development and the dwindling supply of up-to-date material on the Iranian economy. Professor William Polk of the University of Chicago's Center for Middle Eastern Studies encouraged us to move in the direction we had chosen for our inquiry.

We are thankful to Dr. Manuchehr Eghbal, chairman of the board and managing director of the National Iranian Oil Company, who graciously consented to have his staff put at our disposal valuable unpublished documents and research materials. We acknowledge with thanks his staff's cooperation. Dr. Mostafa Mansuri of the Iranian Ministry of Finance and Dr. Ahmad Kooros of the Iranian Central Bank were of valuable assistance to us in furnishing budgetary, oil, and national income data for part of our study. Dr. Freidoon Nasseri double-checked our statistical tables and calculations.

To George Baldwin, a bold explorer of the Iranian economy himself, we owe much for reading the manuscript and for making valuable suggestions. Professor Charles Issawi reviewed the manuscript and made many encouraging remarks for which we are grateful. A special expression of our gratitude, however, should go to Mohammad Yeganeh, who, despite having the heavy burdens of high-level internal and international responsibilities upon his shoulders at the time, cheerfully and tirelessly read every page of the manuscript, using his vast and firsthand knowledge of the Iranian economy, and gave us generous and truly invaluable assistance.

Needless to say, we are solely responsible for any deficiencies and shortcomings of our analysis and synthesis. This study is the result of our own initiative and our own private views. It should in no way be construed to represent the views of Iranian planning or petroleum authorities. Nor should it in any manner be interpreted necessarily to reflect the policies of our government or the responsibilities of our office.

1

Paths to Growth

The world of the late twentieth century is essentially a world of economic development. Political, social, and cultural trends in different parts of the globe, although not always or exactly determined by the behavior of economic factors, are profoundly influenced by them. Most political upheavals, social dislocations, and cultural rebellions are partly rooted in a stream of social consciousness concerning the tempo and direction of economic growth—being too little, too slow, or too lopsided on the one hand, or too independent and disruptive of traditional manners and mores on the other.[1]

Economic development, modestly defined, is a process whereby

---

[1] As early as 1927, Wesley Clair Mitchell ventured that "few problems are more fascinating, more important, or more neglected than the rates at which economic development proceeds in successive generations and in different countries." For several years this observation went unnoticed as the Great Depression of the thirties and World War II occupied economists with more immediate and pressing tasks. But the question of economic development was certain to emerge again and has now been in the forefront of economic thought for more than two decades. See W. C. Mitchell, *Business Cycles: The Problem and Its Setting* (New York: National Bureau of Economic Research, 1927), p. 416.

an economy's real per capita income increases over time.[2] When focus is on the magnitude of this increase, the end product of the development process comes under scrutiny. If we concentrate on the process itself, however, numerous complex relationships and interdependencies will emerge. These relationships are, in general, classified into two groups, according to whether they affect fundamental factor supplies or the structure of demand for products.[3] On the supply side, there are (1) natural resources, (2) capital accumulation, (3) population growth, (4) introduction of new and better techniques, (5) improvement in skills, and (6) other institutional and organizational factors. On the demand side, we have (1) size and composition of population, (2) level and distribution of income, (3) tastes, and (4) other institutional and organizational arrangements.[4] Economic development thus involves a dynamic process with changes in demand-supply relationships.

Our study is a modest inquiry into these relationships in *dualistic* situations—situations characterized by an economically dominant and highly capital-intensive subsector coexisting with a large, traditional, labor-abundant indigenous sector. We use Iran as a case for our study. This choice is fortunate because of the length of time dualistic conditions have existed in Iran and also because the Iranian experience demonstrates the relevance and, we hope, the validity of our approach and our model.

Development Strategies

Economic development has in the past generally been led by the so-called growth centers or leading sectors.[5] The growth sector in some cases has been no more than a single industry—cotton textiles in the British "takeoff" from 1783 to 1803 or railroads in France

---

[2] Some economists have drawn very sharp distinctions between economic "growth," "development," and "progress," trying to distinguish the "quantity" from the "quality" of life brought about by advances in economic indicators. Although conceding the importance of such distinctions, our analysis refers to these concepts synonymously.

[3] See T. W. Schultz, *The Economic Organization of Agriculture* (New York: McGraw-Hill, 1953), p. 5.

[4] See G. M. Meier and R. E. Baldwin, *Economic Development: Theory, History, Policy* (New York: John Wiley, 1963), pp. 2–3.

[5] See W. W. Rostow, *The Stages of Economic Growth* (Cambridge: Harvard University Press, 1962).

from 1830 to 1860. The evolution from maturity to high mass consumption in the United States is believed to have been led by the automobile industry.[6] A leading sector is one that experiences a high growth rate in relation to the rest of the economy and consequently induces favorable response from the other sectors. The inducement is provided by the supply of low-cost products for the consumer, or by stimulating output in other industries, or by creating external economies that can be captured and integrated into other sectors or industries.

There are, broadly, two conventional strategies to development: the "incremental pull," and the "big push." Incremental strategy focuses on marginal movements of resources within the broad framework of *private* supply-demand relationships; it relies essentially on free markets and the price system. The "big push" strategy favors government intervention in the allocation of resources, and is essentially based on the belief in the unworkability of the price system as a means of rapid economic development.

The case against the incremental approach to the development of the emerging nations rests on the contention that the price system, at least in underdeveloped countries, does not work; or if it does, it usually gives the wrong signals. The market mechanism, it is argued, may work well where: (1) the economy is sufficiently monetized and reasonably competitive, so that prices can be held as a fair measure of demand-supply conditions; (2) changes in the supply-demand relationships are only incremental, taking place at the margin; that is to say no large movements of resources in a "big push" are called for; (3) supply-demand elasticities are such that no major indivisibility or bottleneck can block or undermine responses to a change in price ratios; and (4) neither internal nor external economies are of an immense magnitude. In developing countries, where often a sizeable portion of the economy is nonmonetized, where large-scale movements of factors are frequently necessary to break the vicious circles of poverty and productivity, where bottlenecks are enormous, and where externalities are abundant, the price mechanism usually fails to stimulate growth.[7]

---

[6] C. P. Kindleberger, *Economic Development* (New York: McGraw-Hill, 1965), p. 209.

[7] For a fuller discussion of these arguments see, among other works, Hla Myint, *The Economics of the Developing Countries* (New York: Frederick A. Praeger, 1964), particularly chaps. 7 and 8.

In the absence of an automatic mechanism to stimulate growth in the developing countries, the need for deliberate social action is strongly felt. In a pioneering article, some years ago, Rosenstein-Rodan [8] argued that the government must provide the social overhead capital, despite its high capital-output ratio, in order to provide private enterprise with sufficient stimulus to proceed on its own. This plausible view that social overhead capital must precede *private* activity was generally accepted until Hirschman challenged it. [9] In Hirschman's view, it made little difference whether social overhead capital preceded private investment or whether private investment took precedence, thereby creating decision-inducing imbalances. To him, imbalances are corrected through profit opportunities which guide the private investor to the right expansion path. These profit opportunities are created through the so-called backward and forward linkages. A backward linkage is a profit opportunity created for a firm as a result of the expansion of demand for various raw materials and intermediary goods and services generated in the leading sector. A forward linkage is a profit opportunity created for other firms or industries to produce goods and services with the aid of lower-cost inputs produced by the advancing industry or sector. These reactions and unbalanced protrusions are, in short, assumed to lead to further reactions and thrusts, thereby resulting in growth without comprehensive or balanced planning. The primary reason underlying this approach is the acute scarcity of decision-making capacity in underdeveloped countries—a view that has been echoed by many economists. The "big push" approach is designed to maximize the contribution per unit of this scarce resource. [10]

The objective of a "big push" strategy is thus to maintain, in a deliberate and conscious effort, a desirable growth pattern among

---

[8] P. N. Rosenstein-Rodan, "Problems of Industrialization of Eastern and Southeastern Europe," *Economic Journal* (June–September 1943), pp. 202–11.

[9] Albert O. Hirschman, *The Strategy of Economic Development* (New Haven: Yale University Press, 1959).

[10] The argument in favor of incremental development is, of course, based on the assumption that the same output can be produced with varying proportions of social overhead capital and directly productive investment. But if the production isoquants are kinked rather than smoothly curved, social overhead capital must be created before private investment could lead to increased output. To support his thesis for such social overhead planning, Rosenstein-Rodan states in a later article that "neither theology nor technology shows that God created the world convexly downward throughout." See "Planning within the Nation," *Annals of Collective Economy* (1963).

various sectors of the economy. The formulation of such a strategy requires the convergence of many forces, the most important of which is perhaps the role of government. The demand-supply relationship implied in the development process can be brought to bear on the growth of real income more effectively by the correct behavior of the government, as well as by its development strategy. This is of considerable importance in view of the paramount influence that governments can, and do, exert in the development process.[11]

The design of development can, of course, be quite complex, for it involves the tests of comprehensiveness and consistency.[12] Comprehensiveness requires relatively accurate projections of sectoral output levels, input and investment requirements, and the allocation of resources by different sectors for production, consumption, and investment. Consistency requires matching of inputs and outputs in both physical and monetary terms. Since controls are often exercised at the most disaggregated level, comprehensive and consistent planning must provide detailed and precise growth directives. Such multisectoral comprehensiveness and consistency can, in turn, be maintained by estimating the interindustry relations through an input-output accounting system.[13] This system, to be sure, requires a good deal of reliable quantitative data, which is often unavailable or unreliable in underdeveloped countries. However, its use has a number of helpful by-products. It generally forces the planners to state their overall and sectoral targets explicitly. It also provides the planners with a means to explore the consequences of their various alternatives with greater depth.[14]

---

[11] See Albert O. Hirschman, "Economic Policy in Underdeveloped Countries," in *Studies in Economic Development*, ed. Bernard Okun and Richard W. Richardson, pp. 469–76 (New York: Holt, Rinehart and Winston, 1961).

[12] See Raymond Vernon, "Comprehensive Model-Building in the Planning Process: The Less-Developed Economies," *Economic Journal* (March 1966), pp. 57–69.

[13] The output of a certain industry, X_i, must include all its uses as an intermediate product in all industries, X_{ij}, and its final uses—private and government consumption, X_{ic} and X_{ig}, investment, X_{ii}, and exports, X_{ie}. That is:

$$X_i = \sum_{j=1}^{N} X_{ij} + X_{ic} + X_{ig} + X_{ii} + X_{ie}$$

By repeating the same thing for each industry and relating them to future targets, static interindustry balance can be achieved as of a particular date in the future.

[14] For a discussion of input-output accounting methods, see H. B. Chenery and P. G. Clark, *Interindustry Economics* (New York: John Wiley, 1959).

Development Strategy under Dualistic Conditions

In an unbalanced-growth situation, the contribution an industry can make to growth through rapid expansion is likely to vary within a wide range. Depending on the characteristics of demand and the possibilities of forward and backward linkages, the expansion of a certain industry may lead to varying degrees of decision-inducing imbalances. At the same time, the expansion of an industry which may appear to have significant interconnections with other industries may fail to galvanize the other sectors if the latter are incapable of responding to the stimuli, for a variety of reasons including the existence of certain structural bottlenecks.

Direct foreign investment is in general placed in this nonstimulating category. It is argued that such investments commonly constitute an enclave in the underdeveloped country, physically located therein but economically aloof from it as a mere extension of the overseas metropole.[15] Such foreign-based industries as petroleum production, mining, even some manufacturing (e.g., automobiles)[16] that are established in the developing countries with the aid of foreign capital may fail to stimulate the indigenous economy because of low domestic demand and supply elasticities. Such industries frequently become so foreign-oriented that they may obtain their required materials, skilled personnel, and even general labor force from abroad.

A national economy characterized by such foreign enclaves thus comprises two distinct sectors: a highly developed, capital-intensive, usually raw-materials sector, and a relatively underdeveloped, slow-moving indigenous sector. The existence of these two sectors side by side gives rise to what may be described as a "dual economy."[17]

Charles Rollins has made an empirical investigation of the effects of foreign investment in extractive industries under dualistic conditions

---

[15] See H. W. Singer, "The Distribution of Gains between Investing and Borrowing Countries," *American Economic Review* (May 1950), pp. 473–75; and J. Levin, *The Export Economies* (Cambridge: Harvard University Press, 1960).

[16] G. Maxcy and A. Silberston, *The Motor Industry* (London: Allen and Unwin, 1959).

[17] This expression is sometimes also applied to the case of an economy with two distinct "monetized" and "nonmonetized" sectors. In our discussion, "dualism" refers only to the foreign-based vs. the domestic sector.

in Bolivia.[18] He classifies these effects into two categories: *direct* influences and *fiscal* influences. The first includes "all those influences that result from direct contact of the mineral project with the sectors of the economy"; the second includes "those elements that enable the state to contribute to the growth process." Under direct influences, Rollins includes the expansion of markets and incomes resulting from forward and backward linkages and employment of local labor, and foreign exchange receipts placed at the disposal of the country concerned. Under fiscal influences are various governmental activities that could be initiated as a result of the payment of taxes, royalties, etc., by the foreign investor to the host government. These funds could be used to improve market conditions or develop external economies directly (by providing roads and power facilities), or alternatively, by funneling capital into the hands of private persons or enterprises through industrial and agricultural banks that will lend to individuals willing to undertake investment projects.[19]

Rollins concludes that the development of mineral resources in dualistic economies is not likely to lead to general economic growth. His contention is that foreign expenditures within the developing country are rather small in relation to the value of output; the prospects of an accrual of domestic capital in the private sector are rather slight; and the proportion of labor force employed is insignificant because such investments tend to be highly capital intensive. With regard to fiscal influences, the magnitude of direct payments to the host governments may not be very large. Moreover, he argues that the presence of a foreign mineral scheme may well prejudice the possibility of the native government's utilizing funds made available to it by the raw-material industry in order to promote the economic development of that country. Regarding the response of the private sector to the stimulus provided by government policies, Rollins observes that private investors are likely to respond only if the government is capable of directing private expenditures into the desired channels. But such a direction is, in general, contrary to the financial interest of the raw-material producers—at least in the short-run. What Rollins concludes is that foreign investments in extractive industries are not likely to lead to general economic

---

[18] Charles E. Rollins, "Mineral Development and Economic Growth," *Social Research* (Autumn 1956), pp. 253–80.

[19] Theoretically, foreign investors can also put their funds directly into other industries within the host country.

growth, "because they do not become integrated into the economy of the source country to the extent required."

This kind of argument, often echoed with far less sophistication and far more emotional fervor by ardent nationalists in the developing countries, is of much greater portent than is generally conceded in the incremental-pull versus big-push debate.[20] On its face, Rollins's argument denies the applicability of Hirschman's thesis, at least in such countries as Bolivia, Venezuela, or Iran, where the (foreign-oriented) leading sector failed for years to produce growth-inducing linkages. Here are vivid examples of the situations where a significant initial "imbalance" fell short of creating those "pace-setting pressures" of excess demand or supply to put in motion the desired "leapfrogging" effects.

This study intends to argue that Rollins's pessimistic conclusions may have validity only in those cases where (1) domestic political leadership is weak; (2) private-sector response to that leadership is lukewarm; and (3) royalties are meager. That, in fact, foreign investment in strategic industries with a "thick" network of input-output relationships, or under externally propitious circumstances, can indeed produce the forward and backward linkages envisaged by Hirschman. To wit, in an underdeveloped economy of reasonably modest demand and supply elasticities, the indigenous sectors may well respond to the stimulus generated by the dynamism of the foreign-financed operations. The fiscal influences, too, may attain, as they have in some cases, a magnitude sufficient to nurture overall investment and growth. These influences may in turn sufficiently disturb the pattern of supply-demand relationships in the public sector so as to jolt the governments into wide-scale development programs. As will be argued in this study, there is no a priori reason why the response of the government to the disturbance of demand-supply relationships originated by the fiscal influences of foreign investments should always follow Rollins's observations. The specific form and composition of the response of the state depends primarily on its preference function, which may very well look to the right directions.

Given a sufficiently large fiscal influence generated by foreign investment of the type described here, together with an appropriate preference function on the part of the government—as has been

---

[20] See, for example, Keith Griffin, *Underdevelopment in Spanish America: An Interpretation* (London: Allen and Unwin, 1969).

true of Iran in the last one and a half decades—it can be shown that fiscal influences might indeed result in substantial growth via increased public development expenditures. The real problem is the direction of the development effort. Assuming that fiscal influences evoke some systematic, development-oriented responses on the part of the government by means of an internally consistent and comprehensive program, the key to a successful achievement of economic growth will rest with the development strategy.

To be sure, there is no universally valid design for development which could hold under all kinds of situations. But where the development of a raw-material resource (such as oil) is such that it can alter the relative resource structure of the country, a productive development strategy could be aimed at increasing the degree of integration between the foreign-financed, export-oriented sector and the rest of the economy. That is, the development efforts of the government could be geared generally at increasing the forward and backward linkages of such foreign investments by undertaking those investments that would contribute toward raising the relevant demand and supply elasticities. Under these conditions the foreign-financed, export-oriented sector can be expected to contribute to general economic growth via *direct* influences. A well-designed public expenditures scheme under this strategy can, in turn, enhance the productivity of its *indirect* forces.

We believe that Iran presents an excellent case of a developing country which, having been deprived of the beneficial effects of its oil sector for several decades (resembling Rollins's prognostication), has now found a way to make up for the past and to use this dynamic industry as a direct and indirect instrument for its further development.

Our study, designed to bring out some of the important theoretical and empirical aspects of our thesis with regard to Iran, is divided into three parts. Part 1 will inquire into the direct and indirect impact of the oil industry on the Iranian economy during the past sixty years. Part 2 will be devoted to a consideration of some of the fundamental innovative programs initiated and strongly promoted by the Iranian government since 1954—programs which have been directly or indirectly responsible for changes in intersectoral relationships throughout the economy. Part 3 will set forth what we hope is a generalized model of dualistic economies, explaining certain major intersectoral flows of resources and changes in aggregate demand-supply conditions.

Part 1

2

Oil and the Iranian Economy: 1910–50

We begin our inquiry into the impact of oil on the Iranian economy by analyzing the data available for two separate periods: the first period comprises the forty years of oil activities in Iran beginning with the discovery and marketing of oil in commercial quantities around 1910 and culminating in the nationalization of the assets of the former Anglo-Iranian Oil Company (AIOC) early in 1951. The second part will be concerned with the impact of oil on the economy of Iran in the postnationalization period leading to the completion of the Fourth Development Plan in 1972. We devote this chapter primarily to the first period, leaving the discussion of the second period to the next chapter.

Iranian Oil History in a Nutshell

The first successful concession for the exploitation of oil in Iran was granted to William Knox D'Arcy, a British subject, in 1901. Under this concession D'Arcy obtained exclusive rights to the exploration, production, and refining of oil in an area of about 480,000 square miles (all of Iran except the five northern provinces) for sixty years.

In return the Iranian government was to receive £20,000 in cash and an additional sum of £20,000 in paid-up shares of the first company organized to explore oil in Iran.[1] In addition, Iran was to receive a royalty of 16 percent of the company's annual net profits plus all the company's assets upon the expiration of the concession.[2]

It was not until 1908 that a rich strike was made in Masjed-Sulaiman. Immediately thereafter, a new company, the Anglo-Persian (later Anglo-Iranian) Oil Company, with a capital of £2 million, was established and took over all the rights and privileges of the First Exploration Company. In 1914, the British government acquired a controlling interest in the newly formed company by payment of £2.2 million of the new £4 million capitalization.

The original D'Arcy concession was replaced by the "1933 Agreement" which, among other things, (1) reduced the area of the concession to 100,000 square miles; (2) proceeded to settle some of the pending claims of the Iranian government under the D'Arcy concession; (3) reaffirmed Iran's original right to acquire the assets of the company upon the expiration of the contract period; (4) called upon the company to train and employ Iranian personnel for higher-echelon positions; (5) proposed a flat royalty rate of four shillings per ton of oil sold in Iran or exported, replacing the previous royalty rate (16 percent of profits); and (6) provided for the payment of a sum equal to 20 percent of the annual dividends declared by

---

[1] In its dispute with the Anglo-Iranian Oil Company in 1951 the Iranian government claimed that the sum of £20,000 in paid-up shares was initially meant to represent 10 percent ownership in the First Exploration Company. The AIOC claimed that no such percentage was mentioned in the agreement and that the said sum actually amounted to only 3 percent of the capital of the company as organized in 1903. In retrospect, it seems evident that once the concessionnaire got the government to agree to an absolute sum instead of a percentage share he could organize the company with such total capital as to give Iran a very small share. For the text of the agreement see J. C. Hurewitz, *Diplomacy in the Near and Middle East—A Documentary Record: 1535–1914* (Princeton: Van Nostrand, 1956), pp. 249–51.

[2] By the insertion of the word *net* profits in the agreement as the basis for the 16 percent share of Iran, the company also took the liberty of deducting certain items (including income taxes paid to the British treasury) before calculating Iran's share. In retrospect, it seems apparent that the nuances between *net* and *gross* profits were not perfectly clear to the government at the time. Thus, in the Iranian briefs during the dispute, reference was repeatedly made to profit, but the company insisted on the *net* profit. For the views of the Iranian government on the D'Arcy Agreement, see "Iran Presents Its Case for Nationalization," *The Oil Forum* (March 1952), especially pp. 79–83.

the company in excess of £671,250. All together a minimum annual payment of £750,000 was guaranteed by the company.[3]

The terms of the contract, in the view of the Iranian government officials, represented a setback for Iran.[4] The actual outcome of operations was even more disenchanting to the Iranian officials. The ever widening gap between the actual and expected pecuniary and nonpecuniary benefits to the government from AIOC finally reached its limit after World War II, and culminated in a crisis of major proportions.

Although the total annual payment was subsequently raised to over £1 million, and in 1944 to £4 million, the company's royalties fell woefully short of financing Iran's ambitious postwar plans for reconstruction and development. Oil was Iran's only major hope for development finance. So pressure was exerted on the company for a new 50-50 agreement patterned after Venezuela's. But AIOC, fearful of the repercussions of new negotiations on neighboring oil producers and still relying on the British political pressure on Mr. Ala's cabinet, refused to make any major concessions. After a series of long and fruitless discussions between the company and the government, in March of 1951 the Iranian Parliament nationalized the oil industry throughout the country. Shortly afterward, the company's facilities were seized by the government and oil operations came to a temporary standstill.

The result of the 1951 crisis was an agreement in October 1954 between the Iranian government and the National Iranian Oil Company, on the one hand, and a consortium of eight international oil companies on the other.

The Impact of Oil on Domestic Economy: 1910–50

As was indicated in chapter 1, the operation of the Anglo-Iranian Oil Company in Iran before oil nationalization theoretically could have two major effects on the domestic economy: (1) indirect influences via royalty payments to the Iranian government; and (2)

---

[3] For the text of the agreement see J. C. Hurewitz, *Diplomacy in the Near and Middle East—A Documentary Record: 1914–1956* (Princeton: Van Nostrand, 1956), pp. 188–96.

[4] For the views of the Iranian government on the "1933 Agreement," see "Iran Presents Its Case for Nationalization," pp. 83–86.

direct impact through input and output complementarities with other industries in the indigenous sector. We shall first present indirect or fiscal influences on the economy and then shall discuss the interactions between the foreign-financed oil sector and the other "traditional" sectors.

Indirect or Fiscal Impact

During the first half of the twentieth century the oil industry in Iran made continuous progress, with a minor interruption during World War II. By 1950, Iran had become the fourth largest producer of oil in the world, with 6.1 percent of crude oil output, 4.5 percent of refining capacity, and 12.6 percent of the world's proved reserves. Exports of crude and refined oil from Iran during the period amounted to approximately £1,200 million, of which about 10 percent (£120 million) was paid to the government of Iran in the form of royalties, taxes, and share of profits (table 2.1).

From 1911 to 1919 royalties paid to Iran amounted to only about £335,000; from 1920 to 1930 about £10.5 million was Iran's total share; from 1931 to 1940 approximately £26.9 million was received by the Iranian government; and from 1941 to 1951, a total of about £82 million. Despite the almost continuous increase in the production of oil, the proportion of annual average payments to Iran fluctuated between 7.8 and 17.1 percent of the annual average value of exports. It amounted to about 10 percent of the value of oil exports for the entire period. Thus only a very small portion of this lucrative export product accrued to the producing country.

The fiscal influences of oil royalties on the Iranian economy were also undermined by the company's price, dividend, and reinvestment policies. In its price policy, the company apparently discriminated in favor of its major stockholder—the British government—by agreeing to sell oil at discounts to the British navy. The prices charged to the Admiralty were never made public by the company; so the benefits involved cannot be accurately ascertained. Winston Churchill once wrote that the Admiralty contract had saved the British government some £7.5 million during the First World War. On the basis of this estimate, the Iranians claimed that the total savings over the 1910–50 period must have amounted to $500 million—15 percent more than the total oil income received by the Iranian government. The company conceded Churchill's figure but held that the deal was limited to an "exceptional period" during the war. Exceptional or not, it seems plausible to believe that the Admiralty contract hurt Iran both in terms of her share of company profits and in her royalties.

The company also followed a peculiarly low dividend policy throughout most of its life. A major part of undeclared dividends, in turn, went into the expansion of the company's far-flung activities beyond Iranian borders (including the acquisition of a very large tanker fleet). In 1950, for example, of the £422 million gross profit, £142 million was paid out as taxes to the British government, £215 million went into reserves and £45 million to royalties, and £20 million was declared as dividends. The Iranians claim that the company deliberately chose to declare small dividends because, in accordance with the 1933 Agreement, 20 percent of the dividends in excess of £671,250 had to be paid to the Iranian government. The company argued that its policy with regard to dividends and reserves followed "the sound commercial practice pursued by all other companies in large scale operations of an international character."[5] The company's argument can ironically be interpreted to mean that the AIOC was not unique in its fiscal relations with the host country, but that in fact it behaved like all foreign interests exploiting the mineral wealth of poorer and weaker countries.

Iranian officials during the nationalization dispute also claimed that taxes paid to the United Kingdom by the AIOC (in addition to the benefits involved in low oil prices) were in some years more than three times the royalty payments to Iran, which were almost always less than the net profits of the AIOC, after allowing for the royalty payments. According to the Iranian claims, "Up to 1950, the Company and its subsidiaries paid out of earnings to the British Government $1,680 million in taxes and dividends, including the discount to the Admiralty of $500 million. Other stockholders received $170 million, while Iran only received $450 million, or only 9% of total profits."[6]

Owing to the relatively small magnitude of oil income during the 1910–50 period, the budgetary impact of oil was thus correspondingly

---

[5] For the text of the AIOC's brief in the nationalization dispute see "Anglo-Iranian Answers Iran with Facts," *The Oil Forum* (April 1952), special insert.

[6] See "Iran Presents Its Case for Nationalization," especially pp. 82–89. See also United Nations, *Economic Developments in the Middle East, 1945 to 1954*, Supplement to World Economic Report, 1953–54, E/2740ST/ECA/32 (New York, 1955), pp. 70–76; and *Six Decades of Iranian Oil Industry*, undated pamphlet issued by the National Iranian Oil Company.

Part 1

Table 2.1. Net Production, Exports, Sales of Crude Oil in Iran, and Government Revenue, 1911–51

Period	Net Production[a]	Exports[a]	Internal Sales[a]	Iranian Govt. Income (£)	Exchange Rate (Rls./£)	Iranian Govt. Income (Rls.)
9/1/1911–3/31/1912	43,775					
1912/1913	82,097	37,000				
1913/1914	278,026	158,000				
1914/1915	382,010	134,000	6,000			
1915/1916	456,605	172,000	4,000			
1916/1917	654,409	390,000	3,000			
1917/1918	911,803	623,000	3,200			
1918/1919	1,124,170	828,000	2,500	1,325,552[b]	27.70	36,717,790
1919/1920	1,407,531	936,000	2,500	468,718	25.50	11,952,309
1920/1921	1,771,536	1,398,000	5,300	585,290	34.00	19,899,860
1921/1922	2,364,566	2,307,000	5,400	593,429	51.29	30,436,973
1922/1923	3,006,511	2,604,000	5,400	533,251	56.56	30,160,676
1923/1924	3,773,818	3,177,000	8,900	411,322	47.28	19,447,304
1924/1925	4,403,480	3,577,099	11,011	830,754	42.00	34,891,668
1925/1926	4,629,270	4,052,123	12,089	1,053,929	43.50	45,845,912
1926/1927	4,909,336	4,518,505	11,229	1,400,269	48.60	68,053,073
1927/1928	5,443,777	4,754,468	30,705	502,080	49.25	24,727,440
1928 (9 months)	4,358,570	3,781,962	19,227	529,085	49.25	26,057,436
1929	5,548,587	5,416,305	26,576	1,436,764	48.03	69,007,775
1930	6,034,610	5,737,592	25,211	1,288,312	58.00	74,722,096
1931	5,842,776	5,539,203	28,400	1,339,132	89.00	113,182,748

Year						
1932	6,549,244	6,006,298	35,212	1,525,383	100.00	152,538,300
1933	7,200,426	6,654,699	60,603	1,785,013	103.00	183,856,339
1934	7,658,324	7,062,500	83,205	2,159,143	80.24	173,249,634
1935	7,607,852	6,917,768	96,306	2,191,952	79.21	173,624,518
1936	8,329,674	7,244,569	127,719	2,828,502	82.96	234,652,526
1937	10,329,842	9,302,064	153,243	3,444,439	80.00	275,555,120
1938	10,358,976	9,090,052	164,742	3,307,478	80.00	264,598,240
1939	9,731,590	8,020,132	235,193	4,270,814[c]	78.93	337,351,598
1940	8,752,721		278,215	4,000,000	64.83	253,320,000
1941	6,701,274	5,239,787	269,324	4,000,000	85.29	341,160,000
1942	9,546,724	8,030,232	387,808	4,000,000	132.00	528,000,000
1943	9,861,517	8,304,694	716,472	4,000,000	128.00	512,000,000
1944	13,487,255	10,440,050	1,266,391	4,463,778	128.00	571,363,584
1945	17,109,713	14,177,283	842,423	5,623,161	128.00	719,764,608
1946	19,497,486	17,919,178	498,992	7,130,250	128.00	912,672,000
1947	20,518,905	17,725,155	603,537	7,103,795	128.00	909,285,760
1948	25,270,164	22,737,438	701,458	9,172,245	128.00	1,174,047,360
1949	27,236,729	24,632,589	828,172	13,489,271	121.57	1,639,890,675
1950	32,259,642	29,274,187	884,186	16,035,567	89.40	1,433,579,690
1951	16,885,764	14,032,337	907,278	7,000,000	89.40	625,800,000
Total	332,321,085	291,160,129	9,351,127	119,828,678		12,033,413,012

SOURCE: Data supplied by the National Iranian Oil Company.

a In metric tons.

b Includes approximately £335,000 paid to Iran during the 1911–19 period.

c In 1940, an indemnity of £1,500,000 was paid to Iran for 1939 and a minimum payment of £4,000,000 was guaranteed by the company for the years 1940 and 1941, which was later extended to 1942 and 1943.

small. Iran's budget data for the period are too fragmentary to allow a thorough analysis of the budgetary significance of oil revenues. Up to 1927 all the receipts from oil (approximately £7.2 million) were included in the so-called general budget; from 1927 to 1942 only about 6 percent of total receipts from oil (£2.2 million) was taken to the general budget; the remaining 94 percent was earmarked for nonroutine expenditures in special reserve accounts outside the general budget. A comparison between the actual oil payments of the AIOC and those of budget estimates indicates wide discrepancies between the two sets of data (table 2.2). This is partly because in

Table 2.2. Actual Oil Revenue and Estimates of General Budget, 1941–49 (Million Rials)

Year	Actual Oil Revenue	Oil Revenue as per Budget Estimates
1941	341	n.a.
1942	528	347
1943	512	n.a.
1944	571	n.a.
1945	720	512
1946	913	677
1947	909	677
1948	1,174	771
1949	1,640	901

SOURCE: Table 2.1 and United Nations, *Public Finance Information Papers: Iran*, p. 33.

some years only that part of oil revenues that was converted into rials was shown in the budget and the remaining sterling balances were carried in the reserve accounts.

The lack of reliable budget data and the unavailability of information on government reserve accounts make it almost impossible to gauge the precise impact of oil revenues on government budgetary operations. Nevertheless, some broad and admittedly rough observations based on available evidence can be made. Before 1927 almost all receipts from oil were earmarked for current expenditures. But during the 1930s development expenditures became increasingly important, accounting for between 30 and 40 percent of total public outlays[7] (see table 2.3). During and immediately following World War II, however, public development expenditures suffered a heavy setback owing to wartime inflationary pressures. These outlays fell to 18 percent in 1942 and to 7 percent in 1945, and reached only 12 percent by 1947. Included in these expenditures were the purchases of arms and munitions, the construction of the trans-Iranian railway network, and the establishment of some government enterprises. From 1942 to 1948 the oil revenues were again included in the general budget, and from 1949 onward they were divided between the general budget and a newly established development budget.

It is not easy, however, to determine from the budget figures the overall contribution of oil to total government expenditures. Despite the relative importance of development outlays in total government expenditures in certain years, oil revenues constituted only a modest proportion of total government receipts. According to budget data, at no time during the 1910–50 period did royalty payments exceed 15 percent of total government revenues. Although actual budget data are not available, the foregoing estimates provide a fairly obvious measure of the importance of the magnitudes involved. In 1950, the year of the highest annual receipt before nationalization, oil royalties reportedly accounted for about 12 percent of total government revenue and approximately 4 percent of national income.

As these data suggest, the major part of government finances during this period came from sources other than oil. Thus about two-thirds of total government revenues were generally derived from taxes, including special excise taxes on sugar and tea, customs, and road levies. The traditional consumption taxes were relied upon to provide the major part of government revenues. Monopoly revenues, excises, and customs duties on the average furnished about 60 percent of total government receipts in 1937, 1942, and 1948.[8] The relevant data for the years for which they are available are summarized in table 2.4.

---

[7] United Nations, *Public Finance Information Papers: Iran*, ST/ECA/Ser. A/14 (New York, 1951), p. 21.

[8] It may be noted that budget forecasts for receipts have generally been less reliable than those for expenditures. Whereas actual expenditures could not exceed the original appropriations voted by Parliament, the revenue items could, and often did, fall short of expectations. But since actual expenditures were also restricted by administrative restraints, budget deficits did not mount.

Table 2.3. Total Iranian Government Expenditure by Major Components, 1937–49 (Million Rials)

Component	1937	1942	1945	1946	1947	1948	1949
Defense	319	700	1,096	1,545	1,480	1,651	2,478
Administration and others	479	1,168	2,275	3,112	4,561	3,085	4,989
Social services (health and education)	106	231	535	686	777	891	1,232
Interest	7	52	93	124	93	80	50
Capital and development expenditure	662	512	313	390	991	1,197	1,638
Net results of public undertakings	65	—	—	138	119	—	300
Total expenditure (adjusted)	1,638	2,663	4,312	5,995	8,021	6,904	10,687
Debt redemption	—	100	100	100	100	250	430
Total	1,638	2,763	4,412	6,095	8,121	7,154	11,117

SOURCE: United Nations, *Public Finance Information Papers: Iran*, p. 25.

Although oil incomes did not constitute a substantial part of total government revenues, they nevertheless contributed a major share of the country's total foreign exchange receipts. The direct payments to the government, plus the sales of foreign exchange for local currency expenditures, provided valuable foreign exchange to the Iranian economy. In the 1947–50 period, for example, average annual oil royalties, plus sterling received in payment for local currency, accounted for more than 66 percent of Iran's $154 million average annual foreign-exchange earnings.

Direct Influences

By the end of 1950, the Anglo-Iranian Oil Company had a fixed investment in Iran of about £90 million (£28.7 million net of depreciation) with a market value of around £200 million. More than four-fifths of this amount, however, was the result of reinvestment from the company's profits and not from fresh equity capital. The company had discovered approximately 13 billion barrels of proved crude reserves, drilled 453 wells, laid 2,177 miles of pipelines, and built three major ports and 2,500 kilometers of motor road, a few thousand houses, and a number of schools and hospitals. Yet the actual operations of the company were such as to substantially insulate this large industry from the Iranian economy.[9] During the forty-year period, Iran remained mostly an agricultural country with a few light industries set up by the government. No strategic industries for production or processing (other than oil) developed during the period despite the availability of raw materials, cheap labor, and other relatively favorable circumstances.

In a manner similar to that observed by Rollins in Bolivia, the company's direct influences on the Iranian economy were therefore minimal. The flow of resources from the domestic economy into the oil industry—backward linkages—was, except for a contingent of labor, totally insignificant. There was little or no demand on the part of the company for local capital, domestic goods and services, and native managerial talent. AIOC never issued any securities in Iran, or offered its outstanding issues for sale domestically. There was no domestic borrowing, and no encouragement to the establishment of banking, insurance, and other financial institutions.

---

[9] United Nations, *Economic Developments in the Middle East, 1945 to 1954*, pp. 70–71.

Table 2.4. **Total Iranian Government Revenue Estimates by Major Components, 1937–49** (Million Rials)

Component	1937	1942	1945	1946	1947	1948	1949
A. *Revenue from taxes*							
Taxes on income	130	198	644	504	510	600	900
Inheritance tax	—	2	15	7	7	9	10
Land tax	10	—	—	70	150	180	200
Taxes on buildings	2	—	—	6	30	60	50
Outlay taxes and assimilated:							
Excise taxes, mainly kerosine, gasoline, and alcoholic beverages	180	175	360	309	320	445	550
Customs duties	442	307	457	1,205	1,312	1,911	1,679
Fiscal monopolies (gross)	400	1,150	1,406	1,489	1,466	1,870	2,000
Miscellaneous taxes	—	107	24	24	21	31	66
Total revenue from taxes	1,164	1,939	2,906	3,614	3,816	5,106	5,455
B. *Other revenues*							
Oil payments[a]	206	347	512	677	677	771	901
Net profit of public undertakings:							
Post and telegraph (gross)	26	73	122	121	121	126	130
Government domains	42	150	125	125	131	179	189
Government industries (net)	—	1	500	497	408	438	600

Investment income	50	125	51	79	79	126	128
Others and administration	156	109	196	351	327	408	382
Total nontax revenues	480	805	1,506	1,850	1,743	2,048	2,330
Total government revenues (A plus B)	1,644	2,744	4,412	5,464	5,559	7,154	7,785
Deduct for transfer to development budget	—	—	—	—	—	1,065	—
General budget receipts	1,644	2,744	4,412	5,464	5,559	6,089	7,785

SOURCE: United Nations, *Public Finance Information Papers: Iran*, pp. 31–33.

a Oil royalties were not always included in the general budget before 1942 and were kept outside the budget in the reserve accounts. For 1937 the estimated extraordinary budget was available and the receipts recorded in this budget were merged with those of the general budget. The revenue side of the budget was as follows:

Oil royalties	206 million rials
Sugar and tea monopoly	110 million rials
Road tax duties	80 million rials
	396 million rials

It cannot be determined accurately what portion of the total oil revenue is represented in oil payments above since 1942.

As far as local materials production was concerned, the company again offered little stimulus. According to the terms of its agreement, AIOC was free to import its needs mostly exempt from customs duties and other excise taxes. Taking full advantage of this prerogative, the company chose to purchase almost all its needs—even food and beverages—abroad. There is much truth in the Iranian allegations that much of the company's household needs (e.g., food items) and part of its industrial supplies (e.g., cement) were available locally or could have been produced competitively in Iran.

Even with regard to employment opportunities, the company's operations had marginal effects on overall employment, as well as the managerial cadre, and only moderately significant impact on unskilled industrial labor. Of an estimated national work force of about seven million in 1949, AIOC had a total Iranian contingent of around 50,000, or less than 1 percent. This was also less than one-fourth of the total industrial labor force. Of this small contingent, too, less than 9 percent, according to the company's own figures, were *salaried* employees; that is, among technical, commercial, supervisory, and senior staff. The rest were *wage earners*, that is, unskilled workers. And among salaried personnel, the number of "graded" (higher ranking) Iranian employees was about one-third of the foreign (mostly British) staff. The ratio of nongraded salaried employees, on the other hand, was five Iranians to one foreigner. There was no Iranian assigned to a key managerial position within the company.[10]

With respect to forward linkages—the use of the company's output and end products in other complementary industries—the impact, again, seems to have been meager and inconsequential. In terms of physical output, AIOC evidently took no step in establishing, or assisting in the establishment of, by-product industries normally linked to oil production and refining. There are indications that the company was not at all anxious to develop ancillary industries on which it would have to depend for supply.[11] In the area of money

[10] See International Labour Office, *Labour Conditions in the Oil Industry in Iran* (Geneva, 1950), chaps. 2 and 3. The ILO figures obtained from the company differ, in some years considerably, from those estimated by the Iranian government and others. See table 2.5.

[11] The Iranians allege that the company actually "obstructed the growth of industries which would compete for local labor and thereby raise wages." See "Iran Presents Its Case for Nationalization," p. 93.

Table 2.5. Employment in the Iranian Oil Industry, 1939–50

Year	Iranians	Non-Iranians	Not Specified[a]	Total
1939	15,060	2,723	—	17,783
1940	13,380	2,273	—	15,653
1941	10,986	2,079	—	13,065
1942	11,654	1,803	—	13,457
1943	16,389	2,864	—	19,253
1944	16,485	3,380	—	19,865
1945	21,781	4,030	12,143	37,954
1946	24,889	4,520	12,461	41,870
1947	28,221	4,228	11,065	43,514
1948	29,917	4,306	12,189	46,412
1949	32,011	4,477	16,410	52,898
1950	31,875	4,500	n.a.	n.a.

SOURCE: L. Nahai and C. L. Kimbell, *The Petroleum Industry of Iran* (Washington: U.S. Department of the Interior, 1963), p. 20. The data for 1950 have been taken from "Iran Presents Its Case for Nationalization," p. 86.

[a] Includes contract labor, presumably almost entirely Iranian nationals.

and banking, too, the company kept almost all its foreign-exchange earnings, reserves, and deposits in foreign banks, again having no effect on the development of domestic financial institutions.

There was thus really little forward linkage except for the sale of oil for domestic consumption. But even there no real effort was made to encourage the use of oil for home cooking, heating, and other uses. Up to 1923, in fact, all the oil consumed in Iran (about 29,000 tons in 1923) was supplied by Russia while Iran's own petroleum was flowing abroad in ever increasing quantities. By 1929, when domestic consumption reached 82,000 tons annually, some 70 percent of domestic requirements was still coming from the Soviet Union. After the 1933 Agreement, the distribution of petroleum products in Iran became an AIOC monopoly, and although domestic consumption rapidly increased (reaching 270,000 tons in 1941) prices remained uncharacteristically high—being fixed on the basis of prices in the Romanian port of Constanta! The relatively high price of oil, for a country that was virtually floating on it, was partly responsible for a continued and highly wasteful depletion of valuable Caspian timberland at the hands of wood choppers and charcoal makers.[12]

---

[12] See *Six Decades of Iranian Oil Industry*, p. 3.

The high point of our analysis thus far is that the magnitude of direct influences of the oil industry during the 1910–50 period was, for all practical purposes, negligible, and that the industry remained economically divorced from the rest of the Iranian economy. The only major connecting link between oil and the domestic economy was provided by payments of royalties, taxes, and dividends to the government. These payments, too, were of limited benefit, largely because of their relative order of magnitude in the overall budget. Only owing to the limited scope and magnitude of Iranian nonoil exports and the growing needs of Iran for both civilian and military imports, the supply of foreign exchange in oil royalties and sterling conversion into rials (for the company's domestic expenditures) was of relatively notable help to the Iranian economy.

3

Oil and the Iranian Economy: 1950–72

In the preceding chapter we examined the interaction of the oil industry and the indigenous sector of the Iranian economy during the forty-year period from the establishment of the oil industry in Iran up to the time when the industry was nationalized. The overall conclusion suggested by the available empirical evidence indicated that the combined impact of direct and indirect influences of the Iranian oil industry during the first forty years of operation was not large enough to jolt the economy in the direction of self-sustained growth.

In this chapter we shall carry our empirical inquiry to the post-nationalization period. This period has witnessed the launching of four consecutive development plans in Iran. Accordingly, our discussion of the direct and indirect influences of the oil sector will be conducted within the framework of these development plans. In order to present a better picture of the impact of the oil industry on the Iranian economy we shall examine the relative importance of oil revenues in the government's overall budget and the direct contribution of oil to the economic development of Iran in the following two separate sections. But first, a word about the new oil agreements.

The New Era in Iranian Oil History

On 29 October 1954, three months after the change of government in Iran, the three-year-old Anglo-Iranian dispute was finally settled. Under a complicated scheme which was later approved by the Iranian Parliament, an international consortium of several major and some independent oil companies obtained a twenty-five-year (renewable for another fifteen years) concession to exploit and refine oil in southwestern Iran. The former Anglo-Iranian Oil Company accepted the "principle" of nationalization and conceded to the legal ownership by the National Iranian Oil Company (NIOC) of Iran's oil reserves and assets. The NIOC, in turn, authorized the consortium, of which the Anglo-Iranian Oil Company was a principal member, to carry out exploration and refining and marketing operations on its behalf.

Under the 1954 Oil Agreement, the AIOC relinquished its rights in Iran and turned its shares over to the newly formed consortium[1] in return for $1,000 million to be paid in installments by the members of the consortium and £25 million to be paid by the government of Iran over a ten-year period as a net amount of claims and counter-claims of the two parties. The consortium members formed two companies, the Iranian Oil Participants, Ltd., and the Iranian Oil Services, Ltd., both incorporated and registered under British laws in London. The Iranian Oil Participants, Ltd., then proceeded to form two subsidiaries, Iraanse Aardolie Exploratie en Productie Maatschappij (Iranian Oil Exploration and Producing Company) N.V., and Iraanse Aardolie Raffinage Maatschappij (Iranian Oil Refining Company) N.V., generally known as the Iranian Oil

---

[1] The interest in the new agreement was divided as follows: The British Petroleum Company, Ltd. (the former Anglo-Iranian Oil Co.), 40 percent; Royal Dutch Shell Group, 14 percent; Gulf Oil Corporation, Socony Oil Co., Inc., Standard Oil Co. (New Jersey), Standard Oil Co. of California, the Texas Company, each 8 percent; and Compagnie Française des Petroles, 6 percent. On 29 April 1955, the five United States companies turned over one-eighth of their shares to the so-called Iricon Group of companies, which consists at present of the following companies: American Independent Oil Company, Atlantic Richfield Company, Continental Oil Company, Getty Oil Company, Signal Companies, and the Standard Oil Company (Ohio). The text of the original agreement was published in *Platt's Oilgram News Service*, New York, 5 October 1954. See also the white book published by National Iranian Oil Company entitled *History and Text of Iranian Oil Agreements* (Tehran, 1966), in Persian.

Operating Companies, incorporated under the laws of the Netherlands and duly registered in Iran. These two companies have been given the right "to explore for and produce crude oil and natural gas in a defined area of south Iran known as the Agreement Area, and to refine such crude oil and gas."[2] The operating companies are in charge of exploration, production, and refining only, and the marketing function is carried out by the various trading companies organized by the consortium members and registered in Iran. The trading companies, acting individually and independently of one another, purchase the crude oil produced in the agreement area and either resell it for export or have it refined and sell the products for export. The operating companies charge the full cost of production plus a fee of one shilling per cubic meter of crude oil delivered, and one shilling per cubic meter of crude oil refined. (These fees have recently been converted to a percentage of cost.) The National Iranian Oil Company has the right to buy as much petroleum from the operating companies as it needs for distribution and consumption in Iran at cost plus operating companies' fees. In addition, it is also entitled to take 12.5 percent of crude oil output for the purpose of exporting and developing its own market abroad or to sell it to the trading companies at the applicable posted prices. According to a new understanding reached with the consortium late in the 1960s, Iran is also allowed to export crude oil beyond the 12.5 percent limit to markets not traditionally supplied by the oil majors (i.e., Eastern Europe). The amount of such exports in 1970 was 6 million tons. The profits of the trading companies resulting from the difference between the value of oil sold at the applicable posted prices and the expenses of the operating companies are subject to Iran's income tax, which together with the value of the crude oil reserved for the NIOC (12.5 percent of the total) amounted to 50 percent of the profits.[3]

After the conclusion of the 1954 Oil Agreement, Iran's revenue per barrel of oil quadrupled compared with the prenationalization period. Nevertheless, the 50-50 arrangement was still considered unsatisfactory. The 1957 Oil Law gave the NIOC sole responsibility for the development of the nation's oil resources throughout the

---

[2] Iranian Oil Operating Companies, *Annual Review, 1968* (Tehran, 1969), p. 24.

[3] United Nations, *Economic Developments in the Middle East, 1945 to 1954*, p. 74.

country outside the 1954 agreement area.[4] Immediately following this, two "partnership" agreements were signed with AGIP Mineraria —a subsidiary of the Italian State Oil Corporation, ENI, and Pan American Petroleum Corporation—an affiliate of the Standard Oil Company of Indiana. The new agreements in effect provided for a 75-25 division of the net profits in favor of Iran. Shortly after that, several other agreements with similar profit-sharing ratios were concluded between Iran and foreign oil concerns, under which Iran also received the sum of $185 million in 1964 as down-payment "bonus." A recent contractual arrangement with some French oil interests raises the share of profits accruing to Iran, in effect, to 90 percent of total profits. Under the new contractual arrangements, the foreign oil companies serve as contractors in charge of exploration, production, and marketing of oil on behalf of the National Iranian Oil Company. Most of the capital required for operations is supplied interest free by the foreign oil interests and the fixed assets become the property of the National Iranian Oil Company. For their services, the contractors will receive a "contracting fee" and have the right to purchase part of the oil output at stated terms. Table 5.9 in the appendix to chapter 5 summarizes some of the important features of the oil agreements concluded between Iran and foreign oil companies since 1954.

As a result of a supplemental agreement concluded between the consortium companies and the National Iranian Oil Company in 1964 to treat the 12.5 percent stated payment as an operating expense deductible before arriving at the taxable income, Iran's effective share from the consortium operations was raised to 56.25 percent. At the turn of the decade, according to the provisions of an agreement reached in Tehran on 14 February 1971 between the major oil companies and six OPEC countries in the Middle East (Abu Dhabi, Iran, Iraq, Kuwait, Qatar, and Saudi Arabia), the income tax rate applicable to oil companies was fixed at 55 percent for the five-year period ending 31 December 1975—something which had already been secured by Iran under a previous agreement along with a boost of 9

---

[4] Under the oil law passed by the Iranian Parliament on 31 July 1957, the National Iranian Oil Company was authorized, subject to confirmation by the Parliament, to enter into new arrangements with foreign oil concerns for the development of the nation's oil reserves. This was a reversal of the policy adopted in 1947, which forbade the granting of concessions to foreign oil companies.

cents per barrel in the posted rate of heavy crude oil. Furthermore, the pact included a 33 cents across-the-board rise in all posted prices, in addition to a 5 cents per barrel for gravity adjustments, a further 2.5 percent annual increase for "purchasing power" erosions of foreign currencies, an annual escalation of 5 cents per barrel as a five-year "moratorium" settlement, and a 2 cents per barrel increase for freight disparities.

The agreement also provided for the elimination of all previously allowed OPEC discounts and deductions from posted prices, thereby further increasing the governments' revenues from oil. All together the agreement was expected to boost posted prices by 49 cents a barrel during 1971, 55 cents in 1972, 65 cents in 1973, 73 cents in 1974, and 83 cents in 1975. The resulting increases in the revenues of the member governments from oil during the same years were expected to be 30 cents, 34 cents, 39 cents, 44 cents, and 50 cents respectively. The total increase in Iran's revenues over the five years is expected to be $3,600 million, presumably including revenues from increased production. In return oil companies succeeded in securing certain pledges and guarantees from the governments against leap-frogging of claims, imposition of embargoes, and increasing "short-haul" premium rates above 21.5 cents per barrel.[5]

Indirect Influences of Oil on the Economy: 1950–72

The indirect or fiscal significance of oil for Iran in the last two decades, and particularly in the post-1954 period, will, for the sake of greater clarity of exposition and analytical convenience, be analyzed under two main headings: (1) financial aid to the central government in meeting its regular expenditures for traditional public services (e.g., defense, justice, health, education, and welfare); and (2) special assistance to planning authorities in their development finance.

Budgetary Support

Since the conclusion of the consortium agreement in 1954, the growth of the oil industry has been impressive indeed. In the first full year following nationalization, exports amounted to a mere 17 million

---

[5] For details, see *Petroleum Intelligence Weekly*, special supplement (22 February 1971).

Table 3.1. Foreign Exchange Receipts of Iran, 1954–70 (Million Dollars)

Source	1954	1955	1956	1957	1958	1959	1960	1961	1962	1963	1964	1965	1966	1967	1968	1969	1970[a]
Receipts from petroleum sector[b]	34.4	139.2	181.0	256.0	344.1	335.5	358.7	390.1	437.2	470.8	740.4	612.5	715.8	857.4	958.5	1,182.8	1,382.0
	(15.4)[c]	(44.5)	(43.5)	(50.5)	(59.8)	(52.8)	(53.8)	(56.4)	(69.8)	(73.8)	(79.3)	(64.9)	(64.6)	(57.9)	(52.2)	(54.4)	(54.2)
Oil revenue	22.5	92.5	140.5	207.8	244.9	258.7	285.0	289.9	342.2	388.0	466.5	514.1	608.1	751.6	853.5	1,047.4	1,245.0
	(10.1)	(29.6)	(33.8)	(41.0)	(42.6)	(40.7)	(42.7)	(41.9)	(54.6)	(60.8)	(50.0)	(54.5)	(54.9)	(50.7)	(46.5)	(48.3)	(48.9)
Consortium	22.5	92.5	140.5	207.8	244.9	258.7	285.0	289.9	342.2	388.0	466.5	512.1	591.5	715.3	822.5	998.3	1,185.0
	(10.1)	(29.6)	(33.8)	(41.0)	(42.6)	(40.7)	(42.7)	(41.9)	(54.6)	(60.8)	(50.0)	(54.3)	(53.4)	(48.3)	(44.8)	(46.0)	(46.5)
Others	—	—	—	—	—	—	—	—	—	—	—	2.0	16.6	36.3	31.0	49.1	60.0
												(.2)	(1.5)	(2.4)	(1.7)	(2.3)	(2.4)
Purchases from oil companies	11.9	46.7	40.5	48.2	74.2	76.8	73.7	100.2	95.0	82.8	88.9	93.4	107.7	105.8	105.0	134.4	137.0
	(5.3)	(14.9)	(9.7)	(9.5)	(12.9)	(12.1)	(11.0)	(14.5)	(15.2)	(13.0)	(9.5)	(9.9)	(9.7)	(7.2)	(5.7)	(6.2)	(5.4)
Consortium	11.9	46.7	40.5	48.2	74.2	76.8	70.7	96.6	90.5	80.5	77.3	75.6	81.4	82.5	82.8	89.8	92.0
	(5.3)	(14.9)	(9.7)	(9.5)	(12.9)	(12.1)	(10.6)	(14.0)	(14.5)	(12.6)	(8.3)	(8.0)	(7.3)	(5.6)	(4.5)	(4.1)	(3.6)
Others	—	—	—	—	—	—	3.0	3.6	4.5	2.3	11.6	17.8	26.3	23.3	22.2	44.6	45.0
							(.4)	(.5)	(.7)	(.4)	(1.2)	(1.9)	(2.4)	(1.6)	(1.2)	(2.1)	(1.8)
Bonus from new oil companies	—	—	—	—	25.0	—	—	—	—	—	185.0	5.0	—	—	—	1.0[d]	—
					(4.3)						(19.8)						
Exports	106.8	81.1	105.8	123.7	141.3	159.4	167.9	144.3	126.8	148.0	146.0	209.8	225.0	318.1	366.6	419.7	445.0
	(48.0)	(25.9)	(25.5)	(24.4)	(24.6)	(25.1)	(25.2)	(20.9)	(20.2)	(23.2)	(15.6)	(22.2)	(20.3)	(21.5)	(19.9)	(19.3)	(17.5)
Goods[e]	94.9	70.1	89.6	98.4	86.3	94.7	105.6	88.5	82.1	96.9	88.8	132.0	143.8	197.5	208.1	231.3	245.0
	(42.6)	(22.4)	(21.6)	(19.4)	(15.0)	(14.9)	(15.9)	(12.8)	(13.1)	(15.2)	(9.5)	(14.0)	(13.0)	(13.3)	(11.3)	(10.6)	(9.6)
Services	11.9	11.0	16.2	25.3	55.0	64.7	62.3	55.8	44.7	51.1	57.2	77.8	81.2	120.6	158.5	188.4	200.0
	(5.4)	(3.5)	(3.9)	(5.0)	(9.6)	(10.2)	(9.3)	(8.1)	(7.1)	(8.0)	(6.1)	(8.2)	(7.3)	(8.2)	(8.6)	(8.7)	(7.9)

Capital (short term and long term)	33.9 (15.2)	73.0 (23.4)	91.8 (22.1)	100.5 (19.8)	76.4 (13.3)	99.9 (15.7)	115.3 (17.3)	121.5 (17.6)	54.5 (8.7)	19.0 (3.0)	47.7 (5.1)	121.1 (12.9)	167.2 (15.1)	305.7 (20.6)	512.0 (27.9)	569.8 (26.2)	720.0 (28.3)
Other[f]	47.5 (21.4)	19.2 (6.2)	37.1 (8.9)	26.7 (5.3)	13.4 (2.3)	40.4 (6.4)	24.9 (3.7)	35.8 (5.1)	7.8 (1.3)	—	—	—	—	—	—	—	—
Total	222.6	312.5	415.7	506.9	575.2	635.2	666.8	691.7	626.3	637.8	934.1	943.4	1,108.0	1,481.2	1,837.1	2,172.3	2,547.0

SOURCE: *Bank Markazi Iran Bulletin* (July–August 1963), pp. 240–41; ibid. (Sept.–Oct. 1967), pp. 442–43; ibid. (Jan.–Feb. 1970), pp. 644–45; and Bank Markazi Iran, *Annual Report and Balance Sheet as [of] March 20, 1970* (Tehran, 1970), pp. 102–5, Persian edition.

a Estimated by the authors.
b Excludes incidental receipts on account of changes in payment procedure or in the accounting treatment of expenditures.
c Percent of total.
d Less than 0.5 percent of total.
e Includes foreign exchange actually purchased from exporters and hence is not identical with actual exports.
f Includes foreign aid for the period 1954–61.

cubic meters, for which Iran received about $92.5 million. The uninterrupted development and expansion of oil activities in the next decade resulted in increasing oil exports to such an extent that 1969 exports reached upward of 165 million cubic meters, for which Iran received over $900 million in taxes and royalties.[6] During the 1954–69 period, cumulative gross crude production by the consortium companies amounted to 1.3 billion cubic meters, which resulted in a total payment to Iran of about $6.3 billion (table 3.1).

As significant as the absolute amount of oil income has been to the Iranian government during this period, it still fails to reveal the real worth of this income to the state budget.[7] The true value of oil can be measured only in terms of its *relative* share of public revenues. Table 3.2 presents a summary of Iranian government operations for most of the 1960s. As can be seen from this table, annual oil revenues throughout this period were no less than 45 percent, and as much as 54 percent of current public income. Furthermore, the average annual growth of oil revenues—about 18 percent—outpaced the 15 percent annual rise in nonoil income, thus continually raising the oil share in total current income. Table 3.2 shows the revenue components of the "ordinary" (i.e., current) budget for 1963–70. The share of oil income accruing to the Treasury as budgetary assistance is conspicuous by both its annual magnitude and its growth over time. Second only to customs revenues, which have always been the mainstay of government receipts, oil has often been the largest single source of income for the Treasury, surpassing until recently both direct and indirect taxes and monopoly revenues. Since the funds for the maintenance of many projects already completed by

---

[6] Mansour Sadri, *The Impact of the Petroleum Industry in the Economic Development of Iran*, a paper submitted to the Pakistan Petroleum Symposium, Karachi, January 1969, p. 1; and IOOC, *Annual Review 1969*, pp. 40–46; and *Iran Oil Journal* (February 1970), p. 19. See also National Iranian Oil Company, *Annual Report 1969* (Tehran, 1970). Note that the figures given above include payments (stated payments and income taxes) by the consortium members only, which currently account for about 90 percent of Iran's total oil exports. Payments by other oil companies (SIRIP, IPAC, LAPCO, and IMINICO) and payments on account of local currency purchases are excluded from the above figures.

[7] The Iranian annual budget is divided into three subsidiary budgets: (1) the ordinary or current-account budget, which comprises all current revenues and expenditures; (2) the development or capital budget, which shows essentially public investments in development projects; and (3) the budget of public enterprises. Oil revenues are divided between the ordinary and the capital budgets—recently in a one to four ratio.

Table 3.2. Summary of the Iranian Consolidated Ordinary and Development Budgets, 1963–70 (Billion Rials)

	1963	1964	1965	1966	1967	1968	1969	1970[a]
Current revenue	*60.7*	*69.1*	*92.2*	*96.9*	*107.7*	*128.3*	*151.0*	*184.5*
Oil revenue	27.7	36.4	50.0[b]	47.4	54.0	61.8	76.4[c]	95.8
Share of Treasury	(11.4)	(14.1)	(12.4)	(13.3)	(14.5)	(15.0)	(15.4)[d]	(19.9)
Share of Plan Organization	(16.3)	(22.3)	(37.6)	(34.1)	(39.5)	(46.8)	(61.0)[d]	(75.9)
Nonoil revenue	33.0	32.7	42.2	49.5	53.7	66.5	74.6	88.7
Direct taxes	(5.9)	(5.4)	(8.3)	(9.2)	(10.7)	(13.7)	(17.6)	(25.3)
Indirect taxes	(15.5)	(18.1)	(22.2)	(25.4)	(29.6)	(36.8)	(40.7)	(49.1)
Others	(11.6)	(9.2)	(11.7)	(14.9)	(13.4)	(16.0)	(16.3)	(14.3)
Current expenditure	*48.3*	*53.5*	*63.7*	*71.9*	*84.3*	*110.1*	*125.5*	*158.1*
Capital expenditure	*18.6*[e]	*26.7*[e]	*37.1*	*37.8*	*53.4*	*68.2*	*80.7*	*99.6*
Overall deficit	*6.2*	*11.1*	*8.6*	*12.8*	*30.0*	*50.0*	*55.2*	*73.2*
Financing of deficit								
Banking system	5.1	8.3	5.3	5.0	12.5	15.8	19.0	32.7
Foreign loans	1.1	0.8	2.3	4.7	10.1	25.8	23.2	33.5
Treasury bills and bonds	0	2.0	2.0	5.4	7.4	8.4	13.0	7.0
Others and discrepancies	0	0	−1.0	−2.3	0	0	0	0

SOURCE: Bank Markazi Iran, *Annual Report and Balance Sheet as [of] March 20, 1969* (Tehran, 1969), table 39, p. 86, and table 47, p. 99; Plan Organization, *Budget of Iran, 1349* (Tehran, 1970), in Persian, and additional data provided by Bank Markazi.
a Budget estimates.
b Includes Rls. 10.5 billion oil bonus.
c Includes Rls. 6.3 billion advance payment by the consortium companies.
d Estimated by the authors.
e Includes recurrent development expenditures.

the Plan Organization and transferred to regular government ministries (e.g., education, health, roads) are still paid by the Plan Organization through its share of oil income under the guise of the so-called recurrent development expenditures, the contribution of petroleum royalties to the Treasury's finance is even greater than the figure under "oil" indicates.

The oil revenues accruing to the government have thus provided fiscal authorities with a convenient and readily accessible source of funds, thereby relieving them from resorting to many complex (and, at times, growth-inhibiting) fiscal measures to meet operating and development requirements. This means that (1) the state has been able to keep taxes, both direct and indirect, at relatively low levels until the economy attains a sufficiently high and sustained growth rate without substantially impairing public services, and (2) the development and growth objectives of the country have been met without the necessity of siphoning away private consumption, and without undue hardships to private individuals. To put it differently, thanks to the oil revenue, people in Iran have been able to enjoy a higher standard of living concurrently with the larger investment outlays needed for economic development—a situation somewhat different from the more customary pattern of belt-tightening that is generally supposed to precede rapid growth.

Another salutary impact of the oil industry has been its enormous contribution to the country's foreign-exchange earnings. Table 3.1 shows the magnitude of annual foreign-exchange receipts from oil during the 1954–70 period, and the share of oil in total foreign-exchange earnings. As can be readily seen from this table, the amount of foreign currency (i.e., sterling) supplied by foreign oil companies in the form of royalties and sales for local currency was between 52 percent and 79 percent of total receipts on the combined current and capital accounts during the decade of the sixties. As a percentage of current foreign-exchange earnings, oil receipts accounted for a much larger share.

The contributory significance of the oil industry thus lies in the fact that it has been of enormous assistance in alleviating the constraining influence on development that is ordinarily exerted by the scarcity of foreign exchange.[8] The increased leeway provided

---

[8] In India, for example, which lacks such a natural source of foreign income, the shortage of foreign exchange has been considered "the easiest scarcity to see, and it cannot be put down simply as the superficial manifestation of an

by the relative abundance of foreign exchange has allowed the Iranian government to pursue its development objectives with considerably greater ease and without the usual concern for sudden and unpredictable fluctuations in the balance of international payments. Judging from the statistics of the recent past (table 3.1), it is reassuring to note that the inflow of foreign exchange from oil has been consistently high by conventional standards without displaying the vicissitudes characteristic of single-commodity countries. This has important implications from the point of view of the monetary authorities responsible for managing the country's stock of international reserves. For, as is generally conceded, the relative stability of foreign-exchange income can do away with the necessity of keeping very high reserves: the steadier the flow, the smaller the need for stock.[9] This means that, given the high and dependable rate of growth of reserves, the Iranian monetary authorities have managed with smaller reserve stocks than would be required otherwise. The high opportunity cost associated with holding reserves has thus been largely avoided.

Contribution to Development Finance

Concerted development planning in Iran began in 1946 when a fifty-man commission was set up to study the country's resource and growth potential and to formulate a general development program. The Morrison Knudsen Company—a San Francisco-based consulting firm—was commissioned by the Iranian government to study Iran's development possibilities and to submit recommendations to the government. These recommendations and those of the fifty-man commission were further reviewed by a Supreme Planning Board, and finally a skeleton plan with a total proposed expenditure of Rls. 21 billion (nearly $656 million) was prepared.[10] In February

---

underlying inability to save enough or produce enough output." See John P. Lewis, *Quiet Crisis in India* (Washington, D.C.: The Brookings Institution, 1962), p. 38.

[9] J. Marcus Fleming, *Toward Assessing the Need for International Reserves*, Princeton Essays in International Finance, no. 58 (Princeton: International Finance Section, 1967), p. 5.

[10] The expenditure program originally proposed by the fifty-man commission amounted to Rls. 62 billion ($1,937 million). This was reduced by about two-thirds when account was taken of the skilled manpower and financial constraints. It finally was raised to Rls. 26.3 billion ($822 million). Note: the rate of exchange used here is Rls. 32 for one U.S. dollar.

1949, the Parliament passed a Plan Organization Act establishing the Plan Organization for the task of implementing the First Seven-year Plan.

The Rls. 21 billion total expenditure of the First Plan was to be financed and allocated among various programs as shown in table 3.3. The crucial role of oil in the provision of the First Plan's financing

Table 3.3. Estimates of Sources and Uses of Funds for the First Seven-year Plan, 1949–55 (Billion Rials)

	Amount	Percent
Sources:		
Oil revenues	7.80	37.1
Liquidation of government assets	1.00	4.8
Participation by private organizations	1.00	4.8
Bank Melli loan	4.50	21.4
IBRD loan	6.70	31.9
Total	21.00	100.0
Uses:		
Agriculture	5.25	25.0
Roads, railroads, ports, and airports	5.00	23.8
Industry and mines	3.00	14.3
Oil industry	1.00	4.8
Communications	0.75	3.6
Social projects	6.00	28.5
Total	21.00	100.0

SOURCE: Plan Organization, *Report on the Second Seven-year Development Plan* (Tehran, 1343 [1964]), in Persian.
NOTE: Late in 1952, total funds allotted to the plan were increased to upward of Rls. 26 billion.

is abundantly clear from the data in this table. Thus when the nationalization of the oil industry in March 1951 led to a drastic curtailment of oil operations and the incoming revenues, development operations were cut to the bone and the First Plan was altogether crippled.

The financial hardships which gripped the Iranian economy made it difficult for the government to obtain funds from alternative sources of credit. No satisfactory agreement with the International Bank for Reconstruction and Development could be reached (as originally expected), and no credit of any significance was made available by the private sector.[11] The operations of the Plan Organization during the period between the oil nationalization in the spring of 1951 and the resumption of large-scale oil operations in the fall of 1954 was confined mainly to maintaining itself as a "going concern" and moving slowly and inactively along an already beaten path.

With the resumption of oil operations and the increase of oil revenues, the Plan Organization began to step up its activities. Under the authorizations granted in the Development Act of 1949, a number of short-term "impact projects" were formulated to be carried out in the one-year remaining term of the original plan. A new planning board was organized in the meantime to revise the unspent and unobligated authorizations of the original act and to prepare a revised plan for a new seven-year period.[12]

Owing to the substantial interruption of the oil flow during most of the Seven-year Plan period (1949–55) Iran's oil revenues amounted to a mere £76 million or an average of £10.8 million a year. In spite of its financial and other handicaps the Plan Organization made noteworthy strides in different fields of economic development. In these endeavors, the Iranian government was aided by the United States technical assistance program under Point IV and the Mutual Security legislation. In agriculture, the area under cotton, tea, and sugar-beet cultivation was expanded. A livestock laboratory was initiated, pests and animal diseases were fought; farm machinery was imported; and a mobile workshop for repair of agricultural equipment was established. Construction began on Karkheh, Karaj, and several other dams for irrigation, hydroelectric power, and increased water supply; the Kuhrang Tunnel in Isfahan was completed, deep wells were drilled, and water systems were constructed.

---

[11] It may be noted that in order to secure the 32 percent of the funds projected under the proposed plan from the International Bank for Reconstruction and Development, the Overseas Consultants, Inc.—a New York-based consulting firm—was commissioned to study and prepare detailed projects for submission to the bank. A comprehensive report was delivered to the Iranian government after about eight months. However, negotiations between the bank representatives and government officials had barely gotten under way when they were interrupted indefinitely by the deteriorating oil crisis.

[12] For a more detailed exposition of these and other related topics, see George B. Baldwin, *Planning and Development in Iran* (Baltimore: Johns Hopkins Press, 1967), chaps. 2 and 3.

Industry received considerable assistance; two cement factories and four sugar refineries were erected or expanded; construction of a dried-fruit processing factory and a pasteurized-milk factory got under way; and several chemical factories were renewed, replaced, or expanded. In health, an effective antimalaria campaign, waged jointly with WHO and the United States Operations Mission, brought noteworthy results. Transportation and communication, too, received attention: 1,500 kilometers of roads and 100 kilometers of an east-west railroad were put under construction; six airports were paved, expanded, or equipped; and port facilities in Khorramshahr and elsewhere were improved. Banks, private industries, and municipalities also received financial assistance. All in all, more than 130 joint projects were drafted and put into execution under the Point IV program, with varying degrees of success.[13]

Actual investment outlays by the Plan Organization during the First Plan period amounted to over Rls. 4 billion, or less than 20 percent of originally planned expenditures. Transportation/communication claimed less than 40 percent of actual investment outlays, followed by industry/mines and agriculture.[14]

The Second Plan

The Second Seven-year Plan, as ratified by the Iranian Parliament in early 1956, called for a total outlay of Rls. 70 billion ($933 million) over a seven-year period from September 1955 to September 1962. Approximately a quarter of this sum was to be used to complete the unfinished projects initiated in the First Plan, and the remaining three-quarters were to be spent for new projects. About a year later, total authorized expenditures were raised by 20 percent to Rls. 84 billion ($1,120 million). In its final version, this sum was to be allocated among different sectors as follows: agriculture and irrigation, 29.88 percent; transportation and communication, 40.48 percent; industry and services, 11.19 percent; and social services, 18.45 percent.

The Second Plan was to be financed by oil revenues accruing to the

---

[13] For a more detailed description of these projects and their significance in Iran's development see Jahangir Amuzegar, *Technical Assistance in Theory and Practice: The Case of Iran* (New York: Frederick A. Praeger, 1966).

[14] Ministry of Economy, *Industrial Guide to Iran* (Tehran, 1968), p. 58.

Iranian government under the 1954 Oil Agreement, and through foreign borrowing repayable out of future oil revenues. Originally, about 80 percent of oil revenues was to be set aside for development purposes each year. However, the increasing current public expenditures and the reduction of foreign financial assistance forced the government to reduce the share of the Plan Organization, first to 60 percent and later to 55 percent and less. As a result of these changes, the amount of oil revenues actually allocated to the Plan Organization during the plan period came to about Rls. 61 billion, and foreign borrowing to approximately Rls. 26 billion. Table 3.4

Table 3.4. Sources and Uses of Funds by the Plan Organization during the Second Plan, 1956–62 (Million Rials)

Sources		Uses	
Cash brought forward from the First Plan	642	*Development payments*	*75,233*
Revenues	*63,996*	Agriculture and irrigation	23,463
Oil income	60,990	Transportation and	
Interest received on loans		communication	29,940
and deposits	191	Industry and mines	8,824
Exchange rate differential[a]	618	Community development	13,006
Grants-in-aid	611	*Other payments*	*16,587*
Miscellaneous revenues	1,586	Administrative and	
Loans and credits	*29,847*	personnel payments	2,939
Loans	25,607	Repayment of loans	7,623
Short-term credits—Bank		Interest and commission	3,673
Markazi Iran	3,670	Payment of taxes to Ministry	
Short-term credits—Bank		of Finance	566
of Industrial Credits	550	Payment of debts from	
Other receipts	20	First Plan	1,654
		Miscellaneous payments	132
		Balance carried forward to the Third Plan	*2,665*
Total receipts	94,485	Total payments	94,485

SOURCE: Plan Organization, *Report on the Second Seven-year Development Plan*, Appendix 18.
a Due to differential between buying and selling rates of foreign exchange.

summarizes the sources and uses of funds of the Plan Organization during the Second Plan period. The total revenues received by the government from oil during the plan period amounted to around

$1,700 million, and the share of the Plan Organization for the 1956–62 period amounted to about 50 percent. This allocation, however, constituted about 65 percent of total sources of funds available to the Plan Organization during the seven-year period of the plan. It is easy to see, again, that without oil income the Second Plan, too, could not have been successfully launched and properly financed.

The Second Plan, like its predecessor, was beset by some lingering administrative difficulties.[15] The uncertainties surrounding the magnitude of available financial resources (particularly foreign loans), the comparative inexperience in large-scale planning, lack of coordination among various government agencies, and other operating hurdles were instrumental in causing some delays and frustrations. Thus, many programs did not hold closely to their original allocations; projects that were started early naturally established themselves as preferred claimants for funds. Among these were the rehabilitation of two large textile plants, the construction of two new ones, and the building of two new cement mills. These projects were at times seriously threatened by the shortage of funds, but were finally completed. By 1961, the government owned four large integrated spinning and weaving mills with a total capacity of 110 million meters of cloth a year, and the two new cement mills successfully provided the cement requirement of the Sefid Rud and Dez dams.

In agriculture and irrigation, three large dams—Karaj, Sefid Rud, and Mohammad Reza Shah Pahlavi—were completed and studies for additional dams were made. In transportation and communication, 2,700 kilometers of paved highway and 2,800 kilometers of secondary roads were constructed. The railway network was extended from Shahrud to Mashad and from Mianeh to Tabriz, and the overall capacity of the system was increased. The annual combined capacity of Khorramshahr and Shahpour ports was expanded from 870,000 to 2 million tons, and most of the major airports were further equipped and expanded.

In industry and services, with significant help from the private sector, attention was focused on textile, cement, and sugar factories. The annual textile-producing capacity of the country (both public and private) was raised from 60 million meters in 1955 to 418 million

---

[15] See Jahangir Amuzegar, "Iran's Economic Planning Once Again," and "Administrative Barriers to Economic Development in Iran," *Middle East Economic Papers*, 1957 and 1958.

meters in 1962, the cement-producing capacity was increased from 82,000 to 1.2 million tons, and the capacity to produce sugar from 85,000 to 217,000 tons a year during the same period. In social services, significant strides were made toward controlling certain diseases (among them, malaria, smallpox, diphtheria, whooping-cough, and tetanus). The streets of some sixty-three towns were paved with asphalt and such basic utilities as power and piped water were supplied to a large number of cities and towns.

Despite all these and other achievements, the first two Seven-year Plans could still hardly be called "plans" in the technical sense of the term. Strictly speaking they were more in the nature of financial allocations. They did not contain physical targets or explicit state-ments regarding the philosophy and strategy underlying the expen-ditures. The first attempt at comprehensive planning came with the Third Plan, and the approach gained greater sophistication with the Fourth. Special mention ought to be made here, however, of remark-able influences which government training programs, planning guidelines, public expenditures on infrastructural projects, and direct aid to private entrepreneurs (out of the "windfall" gains from rial devaluation) during the Second Plan brought to bear on the expansion and promotion of the almost dormant private sector.

The Third Five-year Plan

The Third Five-year Plan initially proposed a total expenditure of Rls. 190 billion. This was subsequently reduced to 145 and then increased first to 200 and then to 230 billion rials. The plan also proposed an average annual increase in aggregate income of 6 per-cent, based on the assumption of a gross domestic capital formation of 18 percent of GNP derived from an assumed capital-output ratio of three to one.[16]

---

[16] It must be noted that the underlying data were no more than merely "realistic assumptions." For example, the capital-output ratio was obtained "on the basis of comparison with other countries" and assumed to be "in the order of 3 to 1." The derived rate of capital formation, too, was determined after surveying the relevant data pertaining to some twenty-four under-developed countries where "the rate of fixed investment in no case, save one, has exceeded 22 percent of the Gross National Product" and that "high rates of fixed investments have not in all cases given rise to high rates of growth." Plan Organization, *Outline of the Third Plan, 1341–1346* (Tehran, 1342 [1963]), pp. 40–41. The basic planning data of other countries have also been extended to other phases of Iranian planning. See Baldwin, *Planning and Development in Iran*, p. 130.

The basic role of the oil sector is reflected in table 3.5. Oil revenues constitute by far the most important source of development funds

Table 3.5. Sources and Uses of Development Funds during the Third Plan, 1963–67 (Billion Rials)

	Amount	Percent
Sources:		
Oil revenues	144.9	62.5
Oil (bonus)	8.4	3.6
Treasury securities	13.8	5.9
Domestic borrowing	30.0	12.9
Foreign borrowing	20.8	9.0
Other	14.1	6.1
Total	232.0 ($3,093 million)	100.0
Uses:		
Development expenditures	204.6	88.2
Repayment of domestic loans	1.3	0.6
Repayment of foreign loans	8.7	3.7
Interest on Treasury securities and domestic loans	3.8	1.6
Interest on foreign loans	5.5	2.4
Administrative expenditures	3.3	1.4
Other	4.8	2.1
Total	232.0 ($3,093 million)	100.0

SOURCE: Plan Organization, *Report on the Performance of the Third Development Plan* (Tehran, 1347 [1968]), table 9, p. 22, in Persian.

(66.1 percent), followed by total domestic financing (18.8 percent), and foreign loans (9 percent). The Rls. 153.3 billion of receipts from oil allocated to the Plan Organization during the Third Plan period represents more than two-thirds of the total income accrued to the government from oil during the same period. It also represents about three-fourths of total disbursements for development expenditures (table 3.5). The planned and actual allocation of development expenditures among the various sectors are assembled in table 3.6. In both sets of data, transport and communication claim the lion's share, followed by agriculture and power and fuel. Industry and

Table 3.6. Planned and Actual Outlays of Funds under the Third Plan by Sectors (Billion Rials)

| | PLANNED | | ACTUAL | | | |
| | | | Allocation | | Disbursement | |
SECTOR	Amount	Percent	Amount	Percent	Amount	Percent
Agriculture	45.0	22.5	47.9	21.5	47.3	23.1
Industry and mines	21.9	10.9	27.3	12.3	17.1	8.4
Power and fuel	27.0	13.5	35.1	15.8	32.0	15.6
Transport and communication	50.0	25.0	57.0	25.6	53.8	26.3
Education	17.9	9.0	17.6	7.9	17.3	8.5
Health	13.9	7.0	13.3	6.0	13.2	6.5
Manpower	8.0	4.0	2.9	1.3	2.8	1.4
Municipal development	8.0	4.0	7.3	3.3	7.2	3.5
Statistics	0.8	0.4	1.6	0.7	1.5	0.7
Housing and construction	7.5	3.7	12.4	5.6	12.2	6.0
Total	200.0	100.0	222.4	100.0	204.6[a]	100.0

SOURCE: Plan Organization, *Outline of the Third Plan 1341–1346*, p. 72, and *Report on the Performance of the Third Development Plan*, p. 23.
[a] Total includes Rls. 200 million paid to the Ministry of Finance on account of continuing development projects.

mines, too, constitute a major share of total planned and actual expenditures.

The performance of the Third Plan was uneven. In agriculture, for example, a number of exogenous variables such as a severe drought, an extremely hard winter fatal to livestock, and the absence of necessary private investment (owing to initial uncertainties created by the land reform program) caused a reduction of the rate of growth of this sector from a projected 4 percent per annum to a realized rate of 2.8 percent per annum. Industry, on the other hand, had a rather sluggish start but gathered significant momentum toward the close of the plan period, which resulted in an average rate of growth equal to 12.7 percent per annum for the industry sector as a whole. The largest share of the industry allocations, as is shown in table 3.7, has gone to new industries. Included in the latter are the

Table 3.7. Approved Third Plan Allocations for Industry and Mines (Million Rials)

Allocation	Amount	Percent
Technical assistance to private investors	557	2.0
Investment in existing government plants	2,094	7.7
Technical assistance to private investors in mines	278	1.0
Investment in new government mines	398	1.5
Investment in new government industries	18,947	69.3
Long-term credits for private investors	5,048	18.5
Total	27,322[a]	100.0

SOURCE: Plan Organization, *Report on the Performance of the Third Development Plan*, p. 82.
[a] Actual disbursement fell considerably short of this amount.

government joint-venture investments with foreign firms for the manufacture of petrochemical products. There have been three such joint-venture agreements: the first with Allied Chemical on a 50-50 basis to manufacture ammonia, urea, sulfur, phosphoric acid, and diammonium phosphate, mostly for export; the second partnership agreement is with B. F. Goodrich on a 74-26 basis for the production of polyvinyl chloride (PVC), detergents, and caustic soda, aimed for domestic consumption; the third is with AMOCO International (an affiliate of Standard Oil of Indiana) on a 50-50 basis for the manufacture of liquefied petroleum gas and sulfur for export.

The new government industries have thus been the major bene-
ficiaries of the Third Plan industrial program, claiming almost 70
percent of the total allocation. The most important of these new
industries is Iran's steel mill (programmed under the Third Plan
and carried into the Fourth) in the vicinity of Isfahan, and the
petrochemicals complex. Both of these industries will considerably
expand the interaction of the oil sector and the rest of the economy
by augmenting the direct influences of the oil sector on the rest of
the economy. One such direct influence is through the increased
consumption of fertilizers. Fertilizer demand in Iran, although still
modest, is growing rapidly at an annual rate of 35 percent and is
expected to pass the half-million-ton mark by 1972 or 1973.[17]

A notable achievement of the Third Plan in connection with the
much-discussed, long-postponed steel mill was the agreement concluded with the USSR in January 1966. Under this agreement, the
Soviet Union has advanced a twelve-year credit of 260 million
rubles ($286 million) to Iran at an annual interest rate of 2.5 percent
to help finance a steel mill, a gas pipeline, and a machine-tool
factory. Iran has agreed to install the Iranian Gas Trunk Line to
supply gas to the USSR for fifteen years, starting in 1970. The
initial price charged by Iran for the sale of gas will be 6 rubles ($6.60)
per 1,000 cubic meters delivered at the border. The future price will
be adjusted according to fluctuations in world markets.[18]

---

[17] See Baghir Mostofi, "The Petrochemical Resources and Potentials of the
Persian Gulf," *Iran Oil Journal* (December 1968), pp. 11–22.

[18] The rate of supply of gas to the Soviet Union will be initially 6 billion
cubic meters per annum, which is planned to rise to 10 billion cubic meters.
Iran will initially take 2 billion cubic meters annually, which will gradually
rise to 6.5 billion cubic meters. Iran's total income from the sale of gas to
the Soviet Union will be about $36 million in the first year, but is expected
to climb to $66 million by 1975. For an expanded discussion of the Soviet-
Iranian Agreement, see Said Naghavi, "Iranian Gas Trunk Line," *Bulletin
of the Iranian Petroleum Institute* (December 1968), pp. 84–89, in Persian.
The pipeline began operations in October 1970, as scheduled, and gas from
the southern oilfields started the 1,100-kilometer journey to the Soviet Union
for the first time. Here note should also be taken of the expanded trade relations
between Iran and the USSR contained in a new fifteen-year agreement concluded on 7 October 1970. This agreement provides, among other things,
for (1) doubling of Iran's natural gas exports to the Soviet Union and the
construction of a new gas line from the Iranian gas fields to the Soviet border,
(2) expansion of steel production in Iran to 4 million tons a year, and (3)
pooling of efforts by the two nations in the field of petrochemicals and for a
joint exploration of petroleum and gas reserves in the Caspian and Sarakhs
regions.

In other areas, notably in transport and communication, significant progress was achieved. The road-building program continued and received substantial assistance from the International Bank for Reconstruction and Development. In the field of education, the main emphasis was laid upon primary education, and substantially in expansion of facilities rather than improvement of quality. This is because less than 40 percent of children of school age were being enrolled in primary schools. The "Literacy Corps" was highly effective in making possible a substantial improvement in the enrollment ratio to well above 60 percent.

Progress in the first two years of the plan was sluggish, primarily because of the continuation of a Western type of recession that began in the late 1950s. But the rapid growth of economic activity toward the end of the plan period not only compensated for the initially slow growth but set new records in industrial output. Total investment for the period amounted to about $5.7 billion or an average annual rate of about 18 percent of GNP (in 1965 prices) with the highest rate of about 20 percent in 1967. Of this total gross investment, some $2.5 billion (43 percent) was public and $3.2 billion (57 percent) was private, together raising the gross national product and per capita income at constant prices by 8.8 percent and 6 percent per annum, respectively. Special attention ought to be drawn, again, to the crucial role of the private sector in the process of Iran's economic development. As can be seen from investment figures, the share of private investment throughout the entire period has surpassed that of public capital outlays.

All together, the Third Plan not only channeled a substantial portion of the indirect (fiscal) influences of the oil sector into development projects, it was also the first Iranian plan to specify the underlying development strategy, the overall development objectives, and some sectoral growth targets. It thus heralded the era of a more comprehensive and internally consistent planning in Iran. Although severely handicapped by the lack of adequate and reliable data on which to base their calculations and projections, the architects of the Third Plan took a bold and fresh step to formulate the overall goals of the national planning effort, and specify how the oil revenues could contribute to achieving those goals.[19]

---

[19] Although not part of the initial plan projects, mention must also be made of the $90-million Tehran Refinery ($150 million when the Ahwaz-Tehran pipe-

The Fourth Five-year Plan

Iran's Fourth Plan, covering the five-year period from 1968 to 1972, is the most comprehensive and most ambitious of all the development plans yet formulated. It calls for a total investment of Rls. 810 billion ($10.8 billion). About 55 percent (or Rls. 443 billion) of this total will be public, and the remaining 45 percent (or Rls. 367 billion) will be private investment. The new plan seeks to increase the real GNP by 9.3 percent per annum. From an initial $6.9 billion in 1967, the GNP (in 1965 prices) is expected to reach $10.9 billion at the end of the plan period, thereby raising GNP per head to $359 compared with $257 at the end of 1967.

In broad terms, the Fourth Plan aims at (1) heavy industrialization in a variety of fields, including steel, aluminum, copper, lead, zinc, petrochemicals, and engineering industries; (2) scientific water preservation and water-resources development; (3) rapid expansion of power supply (for both industry and agriculture) and the construction of a national grid system; (4) utilization of natural gas for domestic consumption as well as export; (5) rural rehabilitation and urban development; (6) decreasing dependence on foreign markets for food and raw materials; (7) export diversification to reduce heavy dependence on oil income; and (8) modernization of production and management techniques, particularly in agriculture.

Some sectoral targets in the new plan are even more telling. In the industrial sector a steel mill with 1.2 million tons capacity will be completed; a 45,000-ton capacity aluminum plant will be erected; and three petrochemical plants that will produce fertilizers, detergents, liquefied gas, and other chemicals will be finished. In transport and communications, more than 2,700 miles of new highways will be completed; port capacity will be almost doubled from the present 4 million tons; eleven of the country's present airports will be expanded and equipped, and six additional ones will be built. In water resources development, a number of small and large dams

laying costs are added) constructed during the plan and inaugurated in 1968. This refinery, which has a capacity of 85,000 barrels per day (with a potential capacity of 95,000), was financed entirely from domestic sources without foreign capital. See P. Mina, "Progress in the Iranian Oil Industry (1964–1968)," *Bulletin of the Iranian Petroleum Institute* (March 1970), pp. 112–24; and Bank Markazi Iran, *Annual Report and Balance Sheet as [of] March 20, 1968* (Tehran, 1968), pp, 158–65. See also "The Modern Tehran Refinery in Its Third Year of Operation," *Iran Oil Journal* (August 1970), pp. 6, 26–27.

will be constructed. In the field of power, consumption of electricity will be increased almost threefold from 4.5 billion kilowatt hours in 1967 to 12 billion at the end of the plan. In agriculture, one million acres of new farmland will be put under cultivation. Education will be provided for almost all urban, and more than half of rural, children of school age. University enrollment will be increased by 60 percent from the 37,500 students at the beginning of the plan; the number of vocational school students will go up almost three times. More than one-quarter of a million new housing units will be constructed, mainly by private investment. In oil and gas, petroleum production will increase almost 80 percent during the plan period; refining capacity will be expanded by 48 percent; utilization of natural gas will reach 23 billion cubic meters, of which 21 billion cubic meters will be exported. In communications, a microwave network for the entire country will be established, and ten new television stations will be added to the ones already in operation.

The contribution of the oil sector to this extraordinary effort via indirect influences hardly needs elaboration. According to the Fourth Plan, Iran expects to receive Rls. 487 billion ($6.5 billion) from the oil sector, of which 80 percent (Rls. 385 billion) will be allocated to the Plan Organization and the remaining 20 percent (Rls. 102 billion) will be absorbed by the government for routine public expenditures. The share of oil revenues allotted to the Plan Organization constitutes more than 60 percent of the total planned expenditures of the Plan Organization (table 3.8). It also constitutes slightly more than 80 percent of the planned development expenditures (including the related administrative costs). The funds set aside for development by the Plan Organization will be allocated among the various activities as shown in table 3.9.

The vital role of oil revenues in financing the plan, and hence in augmenting capital formation and growth, is readily apparent from the data presented above. The revenues from the consortium companies are most crucial not only in terms of their overall share among the revenue-producing companies, but also because of their relatively greater market vicissitudes owing to the larger quantities of output involved. Given these considerations, it is clear that any significant fluctuations in the flow of oil revenues from the consortium (or any lasting shortfall from the almost 17-percent projected growth rate in oil revenues from the consortium) could have far-reaching repercussions on the ability of the planners to meet their development targets.

Table 3.8. Sources and Uses of Funds of Plan Organization during the Fourth Plan Period (Billion Rials)

	Amount	Percent
Sources:		
Oil income (80 percent of total)	385	63.1
Foreign borrowing	150	24.6
Income from gas and petrochemicals	21	3.4
Domestic borrowing	50	8.2
Other	4	0.7
Total	610	100.0
Uses:		
New development projects[a]	480	78.7
Third Plan projects	45	7.4
Gas transmission	5	0.8
Foreign debt repayment	47	7.7
Domestic debt repayment	28	4.6
Administrative and other	5	0.8
Total	610	100.0

SOURCE: Plan Organization, *Fourth National Development Plan 1968–1972* (Tehran, 1968), p. 62.
[a] Includes Rls. 64.9 billion of administrative costs of development projects.

Notwithstanding the initial foreign skepticism regarding Iran's ability to increase its oil income by the rate assumed in the Fourth Plan, the country has so far managed to outdo even the most optimistic estimates. Thanks to the skillful handling of periodic negotiations between NIOC and the consortium, and particularly as a result of the historic success of OPEC in the Tehran negotiations of February 1971, the Iranian government has succeeded in raising its annual oil revenues far beyond the planned level of over \$1 billion thereby eliminating at the outset one of the most disagreeable uncertainties haunting the plan.

The plan also seems to have gotten off to a successful start in other areas. Up to June 1970, some Rls. 162 billion (or about one-third of the original Rls. 480 billion allocation for development expenditures) was reportedly disbursed by the Plan Organization. Thus all available evidence so far indicates that the planners may even exceed the original target expenditures. The largest share of expenditures so far has gone to industries and mines (Rls. 34 billion),

Table 3.9. Allocation of Development Expenditures by Sectors during the Fourth Plan Period (Billion Rials)

Sector	Fixed Investment by Public Sector	Current Development Expenditures	Investment by Private Sector from Plan Organization Sources	Plan Organization Development Budget
Agriculture and animal husbandry	24.0	20.0	21.0	65.0
Industry and mines	84.7	6.0	8.3	99.0
Gas and oil	26.3	—	—	26.3
Water and power	83.7	2.8	—	86.5
Transportation and communication	97.1	3.2	—	100.3
Rural and urban development	15.1	—	1.0	16.1
Housing and construction	20.1	—	2.9	23.0
Health, education, and welfare	24.6	29.1	1.7	55.4
Tourism	3.6	0.2	—	3.8
Statistics and other	1.0	3.6	—	4.6
Total	380.2	64.9	34.9	480.0

SOURCE: Plan Organization, *Fourth National Development Plan, 1968–1972*, p. 63.

mostly for the steel mill and petrochemicals, followed by oil and gas (Rls. 24 billion) and transport/communications projects (Rls. 26 billion). In July of 1970, the government obtained parliamentary approval for the transfer of up to 30 percent of each sector's appropriation to other sectors. The change was necessitated by the addition of three large new projects (an oil pipeline from the south to Tabriz, a national telecommunications network, and the dieselization of the Tabriz-Julfa railroad). As a result of further appropriations, the total Fourth Plan budget reached Rls. 568 billion in 1971.

As a result, industrial output has increased sharply. The rate of growth of industries and mines in the first year of the plan was about 14 percent, in the second year approximately 11 percent. The GNP also increased at an average annual rate of about 11 percent in real terms compared with 9.4 percent projected in the plan.[20] Most of the increase in industrial production has been due to an expansion of the automotive and electrical industries. Protected against foreign competition, the automotive industry has thrived in recent years. It currently accounts for a significant portion of total industrial investment.[21] Equally buoyant in the first two and one-half years of the new plan has been the petrochemical industry. Investment in this industry has been stepped up. Iran's investment in petrochemicals currently stands at about $350 million. An additional $400 million will probably be invested by the government in the next five years to match an equal sum that is expected to come from private sources.[22] According to one spokesman, the capital-absorbing capacity of the Iranian petrochemical industry now seems to be several times the original $200 million.[23] The Shahpour complex was inaugurated in

---

[20] It should be added that during the first year of the plan (1968) the growth rates of all the sectors (with the exception of the domestic value-added in the oil sector) were above the projected rates, whereas during the second year (1969) the sectoral growth rates (except for oil, water and electricity, and services) fell short of their target rates with agriculture and construction registering the sharpest decline.

[21] The number of vehicles (automobiles, trucks, and buses) on the roads in Iran increased by more than five times in the 1962–67 period and reached 213,500 units at the end of 1967. It is expected that the domestic production of vehicles by the end of the Fourth Plan will be in the vicinity of 70,000 units a year and the total number of new vehicles that will be on the roads in 1972 will reach 75,000. Domestic production of vehicles in 1968 amounted to 34,000 units.

[22] *Kayhan*, international edition, 22 February 1970, p. 1.

[23] *Kayhan*, air edition, 5 March 1969, p. 4. This seems to be in agreement with Mostofi's remark that "the Persian Gulf is . . . well endowed with the raw

1970 and is expected to produce a variety of products, including a daily output of 1,500 tons of sulfur, 1,000 tons of ammonia, and other products with a total value of $150,000 per day or $50 million annually. Iran's stake in this industry is expected to double in the next five years. A $20-million expansion project is already planned for the Shahpour complex that will raise the output of the various products by some 50 percent.

The rapid expansion of petrochemicals has a double-barreled effect: first, it will stimulate a wide range of economic activity, including the production of such widely consumed goods as plastics and related products, soap and detergents, shoes, and most important of all, fertilizers. The latter, as has been pointed out before, have obvious and far-reaching implications for agriculture, which is essential to the growth of the economy as a whole. Second (and of greater fundamental importance to our thesis in this study), the flow of products from the dynamic (oil) sector to the rest of the economy is likely to expand considerably, thereby furthering the integration of the two sectors. For this involves the use of oil not merely as one of several materials that aid the production process, but as a raw material for thousands of synthetic articles and for foods.

The Fiscal Role of Oil in Iran's Development: 1950–72

To sum up, oil revenues have so far been a sine qua non of Iran's economic development, and instrumental in obtaining her remarkable rate of growth in recent years. Up to the start of the Fourth Plan in 1968, some $3 billion of oil royalties were allocated to the Plan Organization for investment in various development projects under the first three plans. A trend toward using an increasing share of oil income for further development has also been accentuated as the absorptive capacity of the economy has expanded and other sources of revenue have become available for the routine operations of the government.

The rising of oil revenues by about 18 percent a year between 1963 and 1970 has helped bring the current level of real national savings to more than 20 percent of the gross national product. This has allowed public savings to rise despite an increase in current public expenditures

---

material prerequisites for a broad-based petrochemical industry." See Baghir Mostofi, "Petrochemical Resources," p. 17. See also Alinaghi Alikhani, "Role of Heavy Industries in Economic Development," *Ministry of Economy Monthly Bulletin*, no. 2, 1345 (1966), pp. 1–11, in Persian.

by more than three times and in investment expenditures by nearly four times between 1963 and 1970. The availability of foreign credit, at moderate rates of interest, to finance the overall deficit—also rising rapidly—has been partly due to Iran's impeccable record as a borrower, but partly also to the expected stream of rising future oil income. The rise of oil revenues in the past has consistently outpaced the increase in nonoil revenues, and has been chiefly responsible for keeping current revenue ahead of current expenditures. In 1970 oil income was about 50 percent of total current public revenues and about 12 percent of the GNP.

A word of caution, however, must be presented at this juncture regarding the overall role of oil income in the government's current revenues. The heavy reliance by the government—any government—on an easy income for the conduct of its routine and development operation at the expense of traditional, painful, and unpopular fiscal tools is bound to produce a new element of risk—a risk that is seldom encountered in developing economies not so fortunate as to have a painless income. More specifically, the dependence of fiscal authorities on oil as an easily accessible source of revenue may tend to retard a smooth and gradual development of a tax base sufficiently broad to be closely interwoven with the mainstream of domestic economic activity. The obvious danger implied by such a reliance is the incongruity that may emerge in the long run between the feeble capability of the fiscal apparatus to generate sufficient revenues and the mounting requirements of a developing economy for such public income. For if the development of necessary fiscal tools and the whole tax-collection machinery is long neglected, it is likely not only to deprive the government of additional revenues, but also to limit the number of policy tools available for influencing the level and direction of business activity.

Fortunately for Iran, this particular risk has always been a major concern for the planners. In fact, the compelling necessity of protecting the national interest against this risk (and that of depletion of domestic oil reserves or a fall in the world consumption of petroleum) has been the subject of daily newspaper editorials, public speeches, and official pronouncement.[24] Reasonable steps, too, have been

---

[24] During the oil nationalization crisis, repeated proposals were made by the National Front leaders for creating an "oilless" economy. See Richard Cottam, *Nationalism in Iran* (Pittsburgh: University of Pittsburgh Press, 1964), pp. 201–2.

taken to hedge against such risks; but the adequacy and effectiveness of current public efforts toward such domestic-resources mobilization do not, of course, enjoy a consensus.

During the 1963–70 period, nonoil revenues have increased by about 15 percent a year—as compared with 18 percent for oil income. Although the level of direct taxes is still relatively low (compared with that of the more advanced countries)—being currently about one-fourth of government nonoil revenues and about 3 percent of the nonoil GNP—reform measures toward expanding the tax base, improving tax administration, and streamlining tax collection have been put into effect in recent years. Income-tax receipts under the Income Tax Act of 1967—from both corporations and individuals—have risen substantially. Nonoil revenues in 1970 constituted about the same proportion of the GNP as oil revenues. The basis for further increase of nonoil revenues from both direct and indirect taxation already exists and is being improved. And the planners seem to realize that the risk is always there.

Direct Influences of Oil on the Economy: 1950–72

The direct impact of oil on the Iranian economy may be discussed in terms of the flow of resources between the oil and nonoil sectors. On the one hand, there is the demand of the oil industry for various outputs of the indigenous sector—for example, capital equipment, labor, supplies, and occasional requirements for incidental goods or services. On the other hand, there is the demand of the domestic economy for the products of the oil industry.

In Iran, domestic demand for petroleum products (rather than oil industry demand for local wares) has been the prime force behind closer integration of the two sectors. The reasons are not hard to find. There is a wide use for fuel in a large number of new domestic industries. And in general it takes much longer for the indigenous sector to be able to supply the highly sophisticated requirements of the oil industry, whereas it is always ready to absorb the stream of products flowing from oil. This basic difference, as we shall see later, is the most important single variable in our dual-economy model. And it is, for all practical purposes, the only viable source of direct intersectoral contacts, culminating in a continuous flow of resources between the two sectors.

In this section we shall first examine the demand-induced influences that initiate the flow of resources from the indigenous sector to the oil sector; then we shall inquire into supply-induced influences that amount to a reverse flow of resources from the oil sector to the rest of the economy. The demand induced influences, as the name implies, refer to those forces that are directly associated with the growth of demand in the oil sector resulting from the expansion of operations in that sector; the supply-induced influences, on the other hand, embody those influences that stem from an abundance of low-cost materials produced in the oil sector.[25]

Demand-induced Influences

The demand originating in the oil sector is itself made up of two components. First, there is demand for fixed assets stemming from decisions by oil companies to expand and enlarge the overall production and marketing capacity of the industry. Second, there is a demand for resources to meet the current and routine requirements of the industry for a given level of operations after desired capacity has been attained. These two components are not, of course, mutually exclusive; they may, as they have since the establishment of the Iranian Oil Consortium, operate concurrently.

The classification of the oil industry demand into capital and current expenditures has a special significance for our purpose. Since oil operations are highly capital-intensive and of a highly sophisticated variety, their demand-induced influences for domestic capital goods tend to be negligible, or almost nonexistent. In the Iranian oil industry the recent trends have been toward greater automation, greater degree of capital intensiveness, and smaller labor component per unit of output.[26] These trends have thus resulted

---

[25] Strictly speaking, the foregoing formulation of the problem is incomplete in that it neglects the impact of changes that may originate in the indigenous sector that are largely independent of the forces at work in the oil sector. Although it is statistically difficult, if not impossible, to isolate the influence of the forces originating in each sector alone, one must at least take into account the important factors—e.g., entrepreneurial or government-initiated activity— operating independently in the two sectors. Given the statistical problem, we shall continue to employ demand- and supply-induced influences as defined above, but shall also take into account some of the important factors in the indigenous sector that have contributed to the augmentation of the intersectoral flow of resources.

[26] The Rls. 11.5 billion investment for the development of Kharg and Mahshahr terminals (the so-called CHAM project), for example, is fully equipped with

in reducing demand-induced influences for capital expenditures. And whatever direct influences may have emanated from the demand side must, therefore, be almost wholly attributed to the current, routine operations of the industry. The underlying reasons for the paucity of capital demands of the industry, as has been pointed out before, must be sought in the current inability of the economy to provide the heavy capital equipment required by the oil industry. Thus the consortium's major investment expenditures have had no appreciable effect on domestic economic activity. This, of course, may not continue indefinitely, because as the economy advances and diversifies so will its capability to supply the capital needs of the industry. But once that stage is reached, the economy as a whole will lose many of the characteristics of a dualistic economy.[27]

The most significant demand-induced influences have thus arisen from the current expenditures of the oil industry. These expenditures include wage payments for labor, purchases of supplies, office equipment, furniture and fixtures, payments for some industrial parts available in the domestic economy, and so on. The latter has become increasingly significant since the establishment of the Internal Resources Division in 1959 in the National Iranian Oil Company charged with (1) investigating and keeping under constant review the possibilities of using Iranian contractual services and Iranian materials in oil operations; (2) initiating and administering the procurement of such Iranian services and materials as they are found suitable; and (3) stimulating by means of advice, guaranteed orders, and sometimes by advance payments the available range of Iranian

---

automatic and push-button devices at amost every phase of transporting petroleum and related products to the loading terminals. See Iranian Oil Operating Companies, *Mahshahr: Terminal of Exporting Petroleum Products* (Tehran, 1346 [1967]), in Persian. The increased use of capital-using, labor-saving methods of production in the Iranian oil industry has led management to reduce the size of the labor force engaged in the industry by encouraging early retirement of personnel, closing down certain previously inhabited production centers, and completely reversing the old policy of establishing workers' living quarters in the producing areas.

[27] To the extent that investment decisions necessitate the expansion of such facilities as buildings, or the employment of local labor for installing and preparing the new plants for production, there is bound to be a rise in demand for domestic resources (labor, materials for construction, etc.). This, however, is rather immaterial in relation to aggregate investment expenditures and will, furthermore, be short-lived unless investment activities of the type associated with the physical expansion of facilities (which generally give rise to new demand for construction materials and labor) continue year after year.

materials and services relevant to oil operations.[28] This division has compiled a five-volume industrial directory containing, as of 1969, the names of a large number of firms in various fields of manufacturing (ranging from refrigerators and related component parts to automobiles, pipes, chemicals and paints, furniture, and office equipment). See table 3.10.

Table 3.10. Recent Additions to NIOC Industrial Directory, 1965–68

TYPE OF INDUSTRY	Number of Manufacturing Firms			
	1965	1966	1967	1968
General-purpose machinery	2	4	13	19
Transport equipment	6	10	7	10
Metal tanks	4	1	4	7
Electrical equipment	5	3	9	4
Pipes, faucets, and related parts	1	2	3	9
Construction materials and tools	19	7	20	15
Water control and irrigation tools	4	1	1	5
Paint and chemicals	5	6	5	13
Medical supplies	9	5	8	21
Furniture and office supplies	60	50	76	75
Miscellaneous	11	8	9	7
Total	126	97	155	185

SOURCE: National Iranian Oil Company, Commercial Section, *Annual Report, 1968* (Tehran, 1969), p. 77, in Persian.

Demand for Local Products

The data in table 3.11 confirm our previous contention that the domestic spillover of the investment demand of the oil industry is bound to be insignificant at early stages of development where a broad and diversified industrial base is largely absent. Although at first glance consortium purchases of foreign products from foreign suppliers seem to have declined in recent years, there has been no appreciable shift to the use of domestic products. Instead, the oil companies' purchases of foreign-made goods, imported to Iran and

---

[28] The Internal Resources Division is also in charge of the "Home Ownership Program" to assist the employees in the oil industry to purchase their own homes.

Table 3.11. Value of Purchases by Oil Companies, 1961–68 (Million Rials)

| YEAR | INTERNAL PURCHASES | | | | FOREIGN PURCHASES | | TOTAL | |
| | Domestic Products | | Imported Products | | | | | |
	NIOC	Consortium	NIOC	Consortium	NIOC	Consortium	NIOC	Consortium
1961	57	535	35	72	439	3,640	531	4,247
1962	92	485	36	37	361	3,558	489	4,080
1963	95	438	60	102	357	2,450	512	2,990
1964	106	420	48	2,051	412	3,029	566	5,500
1965	148	493	57	1,411	464	7,527	669	9,431
1966	107	507	61	1,378	736	1,883	904	3,768
1967	153	538	69	2,344	998	1,983	1,220	4,865
1968	188	693	108	1,758	877	1,860	1,173	4,311

SOURCE: National Iranian Oil Company, Commercial Section, *Annual Report, 1968*, pp. 78–79.

in the hands of domestic suppliers, have risen sharply and correspondingly. Their purchases of Iranian-made products fail to demonstrate a definite trend or an impressive gain. It goes without saying that although the purchases of foreign products by the consortium from domestic suppliers provide certain additions to the GNP's "services" sector, their total contributions are bound to be inconsequential from the point of view of their multiplier impact on the domestic economy because of their high leakage rates. A major exception here is a recent decision by the consortium to purchase the bulk of its electrical energy needs from the Iranian power grid system (largely supplied by the Dez facilities).

Table 3.11 clearly shows that whereas the composition of consortium purchases of domestic and foreign goods has remained much the same throughout the 1960s, the NIOC's purchases of domestic products have been decidedly on the rise, stimulated by its own expansion (independently of the consortium) and also owing to constant prodding from the Internal Resources Division. This difference in the use (and support) of home industries, incidentally, introduces an additional institutional constraint in our model—the nationality of ownership. As can be readily seen, the direct impact of the "dynamic" sector on the economy, in addition to technical and economic factors, would also depend on who owns or controls the industry. In other words, a domestic concern, owned by the government or in the hands of private nationals—even with all the characteristics of petroleum's—would in all likelihood be more inclined to use the products of other domestic industries than would a foreign or joint company. Thus the degree of integration of a largely export-oriented industry into the domestic economy depends, among other things, upon the obligation or inclination of the industry leaders to employ domestic resources.

It should also be noted that a good portion of the NIOC's purchases are also peripheral to oil production and refining, as they mostly relate to a whole array of the so-called nonbasic operations. Included in these operations are such diverse activities as health, housing, road building, training, and social services (clubs, swimming pools, etc.), which are financed by the consortium but carried out by NIOC. Housing probably claims the largest share of the non-basic expenditures (about 30 percent), followed by medical services (about 20 percent) and administration (about 15 percent). Although these expenditures do increase aggregate demand and hence the total level of economic activity, they have little relationship to the input requirements of oil production, exploration, or refining; they are not, therefore, likely to lead to an expansion of those industries that have a direct input linkage with oil. Such nonbasic programs as housing, medical services, and recreational facilities in effect amount to an augmentation of the real wages of the workers and can appropriately be considered as part of the national-income-multiplier effects of wage payments. These effects are also moderated to the extent that part of the required supplies are imported from abroad.

The lack of significant input-output relationships between oil industry operations and the nonbasic activities, however, is not the whole reason consortium officials have tried to sever the industry gradually from all nonbasic operations.[29] The argument frequently advanced by the consortium management for curtailing the nonbasic operations is that oil companies have the prime responsibility of producing, refining, and marketing oil and related products, and like other enterprises they should refrain from engaging in such unrelated activities as housing, medical, and social services. This argument, however, neglects both the ubiquitous nature of the oil industry in Southwestern Iran and the moral and social obligations of an extraordinary enterprise to its employees. To some extent, too, the argument runs counter to the trend in large European and American corporations to provide such services on an increasing scale. A simpler and more plausible explanation for the reluctance of the consortium to continue offering such services is the thankless and frustrating nature of these activities. Consortium leaders remember well how the Iranian "case" for nationalization was packed with references to the neglect of these services by the AIOC. And they want to avoid such embarrassments in the future.

---

[29] Studies are already under way for transferring part of the nonbasic operations to municipalities or other appropriate government organizations. In 1968, the Abadan Institute of Technology, established by the AIOC about thirty years ago, amended its charter to set up a board of trustees to take on the task of administering the school, and simultaneously to sever the institute from the consortium. See National Iranian Oil Company, *Annual Report, 1968*, p. 63. The curtailment of nonbasic operations in recent years is also reflected in the decline of the number of persons engaged in such operations from about 12,000 in 1964 to 8,000 in 1969. See Iranian Oil Operating Companies, *Annual Review, 1969*, p. 46.

Demand for Labor

Another potentially important source of demand-induced influences theoretically could come from the utilization of domestic labor supply both as contractors and as employees and workers. Ordinarily this would be expected to provide a strong stimulus for a continuous expansion of demand-induced influences. But, as has been emphasized before, the highly capital-intensive nature of oil operations has virtually eliminated such an outcome as a logical as well as a practical possibility, thereby rendering the direct influences emanating from demand for labor an impotent source of intersectoral flow.

The empirical data assembled in tables 3.12 and 3.13 substantiate this contention. First, the data reveal that the *total* labor force employed in the oil industry has been declining. Second, the decline in labor force seems to have occurred mainly among the blue-collar employees (from about 48,000 in 1955 to 27,000 in 1970), whereas

Table 3.12. Labor Force Employed in the Oil Industry, 1955–70

| | Staff | | | | |
YEAR[a]	Iranian	Overseas	Labor	Contractors[b]	Total
1955	6,867	85	48,222	88	55,262
1956	7,166	480	47,588	7,913	63,147
1957	7,722	585	47,940	7,056	63,303
1958	8,139	693	48,477	4,724	62,033
1959	8,240	781	47,984	4,305	61,310
1960	8,544	838	45,646	3,206	58,234
1961	10,188	847	39,638	1,619	52,292
1962	9,787	711	33,764	1,554	45,816
1963	9,623	583	32,135	662	43,003
1964	9,888	474	31,564	727	42,653
1965	10,349	501	30,732	2,137	43,719
1966	10,740	506	30,213	1,663	43,122
1967	11,659[c]	—	29,426	1,385	42,470
1968	11,995[c]	—	27,449	2,006	41,450
1969	12,295[c]	—	26,498	2,957	41,750
1970	12,547[c]	—	26,952	1,917	41,416

SOURCE: M. Nezam-Mafi, "Role of the Oil Industry in Iran's Economy" in *Compendium of Speeches on the Iranian Oil Industry* (Tehran: National Iranian Oil Company, 1346 [1968]), p. 284, in Persian, and *Iran Oil Journal*, various issues.

[a] End of year and excluding monthly fluctuations.
[b] Including contract employees.
[c] Including overseas.

Table 3.13. Distribution of Labor Force Employed by Iranian Oil Operating Companies, 1958–69

	PRODUCING OPERATIONS		REFINING OPERATIONS		HEAD OFFICE		
YEAR	Iranian	Overseas	Iranian	Overseas	Iranian	Overseas	TOTAL
1958	18,977	223	24,661	258	181	110	44,410
1959	19,050	240	19,894	304	200	151	39,839
1960	17,278	246	17,678	310	241	171	35,924
1961	14,962	263	22,634	346	428	186	38,819
1962	12,749	217	19,302	261	590	163	33,282
1963	12,068	183	18,191	176	611	140	31,369
1964	11,744	181	18,016	112	554	101	30,708
1965	11,388	169	17,353	112	544	91	29,657
1966	10,947	152	16,293	97	659	98	28,246
1967	9,983	143	14,979	88	628	97	25,918
1968	8,896	144	13,319	73	680	104	23,216
1969	7,886	130	11,857	60	840	105	20,878

SOURCE: Iranian Oil Operating Companies, *Annual Review*, various years.
NOTE: Data for 1958–60 exclude personnel in nonbasic operations.

the number of white-collar employees has generally risen. Third, the decline in the size of the work force seems to be directly related to producing and refining operations. The size of the labor force engaged in producing and refining crude petroleum has clearly shrunk during the past decade despite an unprecedented increase in total production. This is partly the result of the increasingly labor-saving policies of the consortium in Iran. It is partly also the aftermath of the attempts to get rid of "surplus" workers taken on in the 1951–53 period. The rise in the volume of activity, on the other hand, has necessitated the increase in the number of white-collar employees to service the handling of a rapidly rising volume of output (table 3.14).

The net effect on wage payments of a shrinking production-labor force and an expanding service-labor force has been rather mixed (table 3.15). Although wage payments were greater in 1969 than in 1958, they fluctuated rather widely (sometimes by as much as one-fourth or even one-third) during the intervening years. Moreover, wage payments as a proportion of total value-added in the industry (i.e., total contribution to GDP) have been somewhat small (less than 10 percent over the 1962–67 period). They have also constituted

Table 3.14. Gross Production of Crude Petroleum in Iran 1954–70
(Million Barrels)

Year	Consortium	NIOC and Others[a]	Total
1954	11	—	11
1955	120	—	120
1956	198	—	198
1957	267	—	267
1958	304	—	304
1959	343	2	345
1960	388	2	390
1961	434	4	438
1962	480	7	487
1963	533	11	544
1964	613	13	626
1965	668	29	697
1966	745	35	780
1967	905	48	953
1968	992	53	1,045
1969	1,134	100	1,234
1970[b]	1,283	121	1,404
Total	9,418	425	9,843

SOURCE: *Iran Oil Journal* (February 1970); Iranian Oil Operating Companies, *Annual Review, 1965* and *Annual Review, 1969*, and data provided by the National Iranian Oil Company. Figures have been rounded.

[a] NIOC's output declined in 1964 owing to a significant fall in the output of the Alborz wells.
[b] Preliminary.

a negligible proportion of gross national product (less than 2 percent over the 1962–67 period). Wage payments, therefore, can hardly be expected to become a major artery of intersectoral flow of resources.[30]

Of relatively greater importance to the development of the domestic economy (and not brought out by the statistics in the accompanying tables) is the one-way flow of skills and know-how associated with the movement of labor out of the oil industry into the indigenous

---

[30] Specialized contractors could, of course, become a small but highly skilled group of sophisticated enclaves around the oil industry. Unfortunately, the published information about contractors affiliated with the Iranian oil industry is too fragmentary to allow even a cursory analysis of its implications within the framework of this study.

**Table 3.15. Wage and Contract Payments by
Iranian Oil Operating Companies, 1958–69**
(Million Pounds Sterling)

Year	Wages, Salaries, and Fringe Benefits	Payments to Contractors (Including Local Purchases)	Total
1958	21.4	5.4	26.8
1959	22.4	6.3	28.7
1960	21.3	8.2	29.5
1961	30.0	6.8	36.8
1962	29.8	4.7	34.5
1963	23.1	8.1	31.2
1964	25.6	7.1	32.7
1965	27.4	6.4	33.8
1966	29.8	10.6	40.4
1967	32.2	9.9	42.1
1968	34.7	14.7	49.4
1969	36.1	20.2	56.3

SOURCE: Iranian Oil Operating Companies, *Annual Review*, various years.

sector. Although no accurate data have yet been compiled to show the impact of the oil industry on the domestic economy via the transfer of technical and managerial skills, it is believed that the oil industry has played an important role in providing trained manpower to the nation's rapidly growing industries and other economic sectors.[31] In this respect, the industry has helped relieve the economy of one of the most constraining factors that have generally entangled and baffled the developing economies, namely, the dearth of skilled workers and supervisors.

Total Impact of Demand

To sum up, the demand-induced direct influences of the oil industry in the postnationalization period do not seem to have contributed significantly to the intersectoral flow of resources. The reasons are not hard to find. First, the basically capital-intensive nature of the oil industry has been unfavorably inclined toward the use of much

---

[31] *Six Decades of Iranian Oil Industry*, p. 15.

labor. Second, the familiar aversion of the consortium, as a foreign concern, to becoming overly involved in the livelihood and welfare of "native" workers has induced its management to seek a refuge in increasing and deliberate automation. Third, the unpreparedness of the Iranian economy to provide the capital equipment requirements of a fast-growing and modern industry like oil (owing to a lack of sophisticated engineering industries in the indigenous sector) has given the consortium a plausible excuse for obtaining its needs from foreign sources or through imported products. And fourth, special privileges obtained by the consortium in the 1954 Oil Agreement have enabled it to flout Iranian government pressures for the use of domestic products.

Supply-induced Influences

Supply-induced influences, as has been noted before, contribute to the intersectoral flow of resources by giving rise to a continuous movement of low-cost raw materials from the dynamic (oil) sector into the domestic economy. The huge quantity of cheap primary materials produced in the leading (oil) sector provides an inducement for the consuming units in the indigenous sector to increase their consumption of those materials by embarking on the production of goods that use the products of the oil industry relatively more intensively. Thus there will be a tendency, slow as it may be initially, on the part of the consuming units in the domestic economy to substitute petroleum products for other types of fuel and move in the direction of using petroleum not only as a minor component of total inputs but as the main raw material put into process.[32]

---

[32] To suggest that there may be a tendency for supply-induced influences to increase the intersectoral flow of resources does not mean that there *will* be an "automatic" flow of materials from the dynamic sector (oil) to the domestic economy as soon as a sufficiently large quantity of low-cost (petroleum products) become available. Such a suggestion would admit a forthright applicability of the Schumpeterian growth theory to the conditions of a dualistic economy. That Schumpeter's theory requires some modifications for the less developed countries has been forcefully argued by Professor Wallich. Suffice it to state at this point, subject to certain qualifications, that the main propelling force in the increase of the flow of resources from the leading sector to the rest of the economy is not always or automatically provided by private entrepreneurs (as is assumed in the Schumpeterian model), but may require government action supported by the public's desire for higher living standards. The substitution of government action for entrepreneurial initiative as a catalytic agent in the

A distinction ought to be made at this point between the relative importance of demand- and supply-induced influences in the development of an underdeveloped economy. In general, forces of both supply and demand have essentially similar functions in predisposing a developing economy toward industrialization. For a leading manufacturing industry producing final goods (e.g., automobiles), demand influences may assume predominance. In mineral and extractive industries (particularly a versatile raw material like oil) the leading sector may provide its dominant influence through supply.

In the case of Iran and its petroleum industry, unlike the secondary role played by demand-induced influences in augmenting the flow of resources between the two main sectors, the supply-induced influences are likely to assume a far more dominant role. Our particular development model sketched below (unlike the Schumpeterian theory that draws its motive power from the sphere of entrepreneurial decisions) is based not so much on innovation and entrepreneurial initiative, but mainly on the absorption of borrowed technology and the leading role of the government in the context of popular demands for better living standards.

The industrialization of the economy will, in our model, generate a demand for certain primary products, which can, in part, be supplied from the oil sector. The relatively low cost of these products will also push the economy toward production of those goods that use the raw materials provided by the oil sector relatively more intensively. As the domestic economy advances, the flow of resources (i.e., primary products) from the oil sector into the rest of the economy is likely to rise.

The issues raised by our analysis are borne out by the data on Iran. Petroleum products account for a major portion of the total energy consumed in Iran. The consumption of various oil products in 1970 amounted to about 185,000 barrels per day (10.8 million cubic meters a year) or about two barrels a year per capita. As late as 1959, the total domestic consumption of oil products was only

---

development process, however, does not alter the end result (i.e., the flow of resources from the oil sector into the rest of the economy); it merely changes the motive force as well as the process by which the end result is achieved. See Henry C. Wallich, "Some Notes towards a Theory of Derived Development," in *The Economics of Underdevelopment*, ed. A. N. Agarwala and S. P. Singh, pp. 189–204 (New York: Oxford University Press, 1958).

3.4 million cubic meters. The average annual growth of domestic demand for petroleum products in Iran has ranged between 12 and 15 percent in recent years, signifying the highest per capita consumption in the ECAFE region, after Japan.[33] National consumption is expected to reach 485,000 barrels per day (28 million cubic meters a year) or about five barrels a year per capita by 1979. Of the different petroleum products consumed, fuel oil has consistently accounted for the highest proportion of total consumption since 1957. Moreover, it has displayed one of the highest rates of growth in recent years, rising from less than 850,000 cubic meters in 1957 to over 2.75 million in 1969. It is expected to accelerate even more over the next few years.[34] This is a significant development indeed, as the consumption of fuel oil is directly linked to the energy requirements of such important domestic industries as sugar, cement, and power.[35] See tables 3.16 and 3.17.

Part of the increase in the total consumption of oil products is due to the facts that (1) kerosine and fuel oil are rapidly replacing wood, charcoal, and animal manure as cooking and heating fuel; (2) gas oil is being substituted for gasoline for use in internal combustion engines because of a substantial price differential; and (3) gasoline use is expanding owing to increasing mechanization of transportation means, particularly motor vehicles. In a study by the Bank Markazi Iran, the income elasticity of demand for both fuel oil and gas oil was computed on the basis of 1959–68 data and was found to be significant. The bank report concluded that the findings were indicative of "the extent of industrialization in the country over the past few years."[36]

Special mention also ought to be made of the consumption of natural gas for residential and commercial purposes. Bottled liquid

---

[33] National Iranian Oil Company, *Annual Report, 1968*, pp. 20–21; and Mansour Sadri, *Impact of the Petroleum Industry*, pp. 3–4.

[34] National Iranian Oil Company, Distribution Department, *Projection of Consumption of Petroleum Products in Iran, 1969–1979* (Tehran, 1969), p. 1, in Persian; and Bank Markazi Iran, *Annual Report and Balance Sheet as [of] March 20, 1969*, pp. 182–83.

[35] These industries are major consumers of fuel oil. For example, half a liter of fuel oil is on the average consumed for the production of one kilogram of sugar, and about 210 liters of fuel oil are consumed to produce one ton of cement. For further details, see *Projection of Consumption of Petroleum Products in Iran, 1969–1979*, pp. 28–29.

[36] *Annual Report and Balance Sheet as [of] March 20, 1969*, p. 182.

Table 3.16. Domestic Consumption of Petroleum Products in Iran, 1932–70 (Thousand Cubic Meters)

Product	1932	1937	1942	1947	1952	1957	1958	1959	1960	1961	1962	1963	1964	1965	1966	1967	1968	1969	1970
Gasoline	20	94	101	164	262	464	524	575	627	644	664	702	740	714	800	854	943	1,068	1,121
Kerosine	9	56	81	162	336	713	752	886	976	1,099	1,171	1,258	1,426	1,470	1,530	1,808	1,990	2,336	2,498
Gas oil	—	13	29	46	124	385	520	707	856	984	1,083	1,158	1,349	1,593	1,881	2,185	2,467	2,654	2,978
Fuel oil	18	28	156	326	494	828	874	1,027	1,147	1,222	1,259	1,268	1,507	1,712	1,979	2,238	2,577	2,820	3,376
Other products						93	136	172	256	309	275	278	355	400	482	602	666	832	803
Total	47	191	367	698	1,216	2,483	2,806	3,367	3,862	4,258	4,452	4,664	5,387	5,889	6,672	7,687	8,643	9,710	10,776

SOURCE: National Iranian Oil Company, Distribution Department, *Annual Report, 1967* (Tehran, 1968), and *Annual Report, 1969* (Tehran, 1970), both in Persian, and Nezam-Mafi, *Role of the Oil Industry*, p. 282.
NOTE: Consumption of petroleum products by the oil industry is excluded; domestic consumption of natural gas is excluded; totals for 1932–52 comprise the four major products only. 1970 data are preliminary actuals.

Table 3.17. Projected Domestic Consumption of Petroleum Products in Iran, 1971–79 (Thousand Cubic Meters)

Product	1971	1972	1973	1974	1975	1976	1977	1978	1979
Gasoline	1,222	1,332	1,452	1,583	1,725	1,880	2,049	2,233	2,434
Kerosine	2,723	2,968	3,235	3,526	3,843	4,189	4,566	4,931	5,325
Gas oil	3,313	3,703	4,215	4,790	5,430	6,157	6,966	7,844	8,848
Fuel oil	4,002	4,829	4,466	5,131	6,017	6,917	7,528	8,390	9,497
Other products	924	1,045	1,164	1,289	1,426	1,569	1,728	1,898	2,085
Total	12,184	13,877	14,532	16,319	18,441	20,712	22,837	25,296	28,189

SOURCE: National Iranian Oil Company, Distribution Department, *Projection of Consumption of Petroleum Products in Iran, 1969–1979.*
NOTE: Projections assume no substitution by natural gas for oil products; consumption of oil products by the oil industry is excluded.

gas has been supplied by the NIOC and marketed by private distrib-
utors since 1955. The phenomenal growth of this "clean" fuel—
from a mere 80 tons in 1955 to over 82,000 tons in 1969—promises
to revolutionize home cooking, heating, and air conditioning across
the country, particularly in urban centers. Consumption is expected
to reach 200,000 tons in 1972, supplying about one million families.
With the completion of the 1,100 kilometer Gachsaran-Saveh-
Astara main gas pipeline to the Soviet Union, households and
industrial plants in the vicinity of the pipeline can now expect to be
supplied with abundant low-priced natural gas. Plans are also under
way to connect the gas main to Tehran, Isfahan, and Shiraz in the
1970s, and to Gazvin, Rasht, and Pahlavi later. Gas consumption
is expected to reach 600 million cubic feet a day by then, almost
half of it for Tehran alone. The conversion from the present liquid
gas plus kerosine to natural gas is expected to save an average
household in Tehran more than 25 percent in fuel costs.[37]

Of special importance among supply-induced influences in recent
years has also been the unprecedented activity in the field of petrochemicals. The locus of Iran's petrochemical activity is the National
Petrochemical Company—a subsidiary of the National Iranian Oil
Company—with a total capital of $110 million. This company
has set up several single and joint ventures that have together given
birth to a fast-growing petrochemical complex in the country. Among
them, as was indicated before, are: the $24 million Iran Fertilizer
Company in Shiraz, with an annual output of 65,000 tons of fertilizers (ammonium nitrate and urea) out of natural gas; the Shahpour
Chemical Company, a 50-50 partnership with the Allied Chemical
Corporation for the production of fertilizers and sulfur, with an
original contributed capital of $25 million; the Abadan Petro-
chemical Company, a 74-26 alliance with B. F. Goodrich, with a
contributed capital of $8 million for the production of PVC, DDB,
and caustic soda; the Kharg Chemical Company, a 50-50 venture
with AMOCO International (an affiliate of Standard Oil of Indiana),
with a contributed capital of $7 million for the production of LPG
and sulfur; and Pazargad Chemical Company for the production
of caustic soda, in which the National Petrochemical Company and

---

[37] In 1969, Shiraz was the only city in Iran using natural gas at the rate of
20 million cubic feet a day. See NIOC, *Seyri dar Sanaate Nafte Iran* (*A Journey
across the Iranian Petroleum Industry*), 1348 [1969], pp. 18–37.

the Consortium Operating Companies own a controlling interest and the remaining stocks are privately held.

Operations have already begun in most of these plants that will produce a wide variety of products for export and internal consumption. This new species of industrial activity has emerged primarily because of the low-cost and abundant supply of hydrocarbon. The internal demand for fertilizers is expanding and is expected to continue to climb rapidly, so that by the end of the 1970s it will be economically feasible to set up and operate additional large plants on the basis of internal demand alone.[38] This is an obvious example of supply-induced influences where a whole new line of activity is beginning to emerge, albeit by direct government action, around a low-cost source of supply.

Thus, the supply-induced influences of the petroleum industry seem to be beginning to exert widespread synergistic effects on the whole economy. And in all likelihood the major direct impact of oil on the Iranian economy is bound to come from supply-induced influences rather than from the demand side. For during the industrialization phase of development, when the economy enters the take-off stage, the availability of low-cost materials in the dynamic (oil) sector would induce the domestic economy to produce those goods that use these materials more intensively. Moreover, the magnitude of the flow of these resources will grow as the domestic economy industrializes further and adjusts to changing supply-demand conditions. On the other hand, given the complicated equipment requirements of the oil industry, the domestic economy may long remain unable to supply the needs of the oil sector, except perhaps for the relatively minor component needs and, of course, labor.

---

[38] Ibid., p. 98.

Part 2

4

Achievements and Drags

An economy's successful performance is not easy to define. Harder still is the measurement of economic success or failure. There are few objective criteria for such evaluations or assessments. If success could be defined as the attainment of desired objects or objectives, then a nation's economic achievements would have to be judged mainly in terms of its own socioeconomic goals, and by the extent to which these goals are obtained. The same approach can be followed in the case of Iran.

Iran's national socioeconomic objectives in its postwar planning have included a high rate of economic growth, price-level stability, overall balance in external accounts, equity in the distribution of income, economic security, and production of certain essential or "strategic" goods and services at home. The government has tried to increase the national income (ranging from 6 percent to 9 percent annually under different plans) by increasing public investments and encouraging private capital formation. The planners have also aimed at keeping price levels and balance-of-payments deficits within reasonable bounds—for political reasons as well as for considerations of social justice. A narrowing of the consumption gap between the rich and the poor through raising the minimum standard of living and redistributing individual incomes has also been a major social goal. Social justice has been expected to be achieved through increased economic security for the new farmers, the urban proletariat, and the civil servants. Economic diversification has been planned through a program of industrialization, national self-sufficiency in certain consumer items, and emphasis on exports of nontraditional products.

Against the backdrop of these socioeconomic goals a cursory appraisal of Iran's economic performance in the postnationalization period—1955 to 1970—can be attempted. The following evaluation —relating mostly to the achievements of the 1960s—is a summary of major national accomplishments in the areas of growth, economic stability, individual welfare, and product diversification. The remainder of the chapter will discuss the resulting structural shifts in the economy. A brief discussion of the development "drags" will bring the chapter to an end.

Growth

During the 1960s Iran has experienced a remarkably high rate of economic growth in actual prices, real terms, and per capita. Output during the decade, starting from a somewhat depressed level initially, has grown steadily at an annual rate of some 7.6 percent in real terms (and 9.5 percent at current prices)—exceeding the original 6 percent growth target of the Third Plan (1963–67), and matching the rates achieved by some of the fastest-growing countries in the world. As population is estimated to have grown between 2.7 and 3 percent per annum during the same period, real per capita GNP has increased by about 4.5 percent a year. In some years (e.g., 1965) the overall growth rate even reached 12 percent in real terms. Some industries, like oil and manufacturing, have had annual increases outpacing the overall growth rate of the economy. In the 1954–60 period for which some rough data are available, the rate of increase is estimated around 4.5 percent at current prices.[1] The Third Plan achievements were also evidently obtained by a capital-output ratio

---

[1] The estimates of output and income for the years before 1960 are mere approximations and should not, therefore, be taken to represent precise changes of corresponding data.

of slightly more than two to one—a low ratio for a developing country.[2] The gross national product in 1969 stood at almost Rls. 700 billion (just over $9 billion) at current prices and Rls. 583 billion (nearly $8 billion) in 1959 prices. Per capita GNP was $330, and per capita income $285, comparing favorably with per capita GNP of $200 and per capita income of $175 in 1962.

All major sectors of the economy have contributed more or less to the overall growth. Oil, growing at an average annual rate of about 15 percent in revenues in the 1960s has accounted for about 33 percent of the growth in total value added during that decade. Industry, advancing at an average annual rate of 11.3 percent has provided another 19 percent of the total added value. Manufacturing in particular has had a very notable growth. Although the traditional sectors—food processing, textiles, construction materials, and carpet weaving—have continued their dominant role, new industries, mainly electrical, chemical, metalworking, and machinery, have expanded at more than 12 percent per annum. In 1969, the automotive industry showed one of the fastest growth rates, contributing more than one-fifth of the total industrial growth.[3] In 1970, the total number of "industrial" units—large factories, small plants and workshops—is estimated at more than 550,000, employing a total labor force of close to 1.5 million.

The major elements of growth have thus been provided by (1) the oil sector, in which local efforts and initiatives have played a relatively minor economic role, and (2) industry (including power, construction, transport, and communications) which has been more amenable to increased investment and production through increases in importation of capital goods. Part of the total annual growth has also been provided by the expansion of the services sector which includes the government's own administrative apparatus. Thus an increase in the number of public employees and public payroll has been a major element in the statistical growth of public services. In 1969, the growth of these services was given as 15.5 percent.

Somewhat disappointing, however, has been the growth behavior of agriculture. During the 1960s, farm production rose by an average annual rate of a mere 3 percent—barely exceeding the annual rise

---

[2] Some inordinately favorable factors (e.g., good weather, unused capacity, and a sudden jump in oil output), however, have helped to keep the ratios down.

[3] *Kayhan*, international edition, 15 June 1970, p. 2.

Table 4.1. Distribution of National Product by Final Use at Constant 1959 Prices, 1959–70 (Billion Rials)

	1959	1960	1961	1962	1963	1964	1965	1966	1967	1968	1969	1970[a]
Consumption expenditure	239.2	244.5	253.7	262.5	273.9	297.7	325.6	346.6	379.0	419.5	462.1	500.0
	(85.3)[b]	(83.4)	(81.5)	(81.5)	(80.8)	(81.6)	(79.7)	(80.0)	(78.3)	(80.2)	(79.3)	(78.7)
Private	209.1	212.7	221.7	230.7	240.1	256.4	272.8	287.6	313.8	333.3	356.5	382.5
	(74.6)	(72.6)	(71.2)	(71.6)	(70.8)	(70.3)	(66.8)	(66.4)	(64.8)	(63.7)	(61.2)	(60.2)
Public	30.1	31.8	32.0	31.8	33.8	41.3	52.8	58.8	65.2	86.1	105.6	117.5
	(10.7)	(10.8)	(10.3)	(9.9)	(10.0)	(11.3)	(12.9)	(13.6)	(13.5)	(16.5)	(18.1)	(18.5)
Gross domestic fixed capital formation	48.1	54.4	55.8	49.3	53.2	61.9	79.3	86.5	108.2	121.7	125.3	135.0
	(17.2)	(18.6)	(17.9)	(15.3)	(15.7)	(17.0)	(19.4)	(20.0)	(22.3)	(23.3)	(21.5)	(21.2)
Private	31.9	38.9	37.4	33.6	35.4	42.6	46.1	51.6	56.1	61.6	62.0[a]	65.0
	(11.4)	(13.3)	(12.0)	(10.4)	(10.4)	(11.7)	(11.3)	(11.9)	(11.6)	(11.8)	(10.6)	(10.2)
Public	16.2	15.5	18.4	15.7	17.8	19.3	33.2	34.9	52.1	60.1	63.3[a]	70.0
	(5.8)	(5.3)	(5.9)	(4.9)	(5.3)	(5.3)	(8.1)	(8.1)	(10.7)	(11.5)	(10.9)	(11.0)
Exports of goods and services	60.2	68.6	74.5	82.5	85.7	90.5	104.4	115.1	136.9	155.0	180.0[a]	205.0
	(21.5)	(23.4)	(23.9)	(25.6)	(25.3)	(24.8)	(25.5)	(26.6)	(28.2)	(29.6)	(30.9)	(32.3)
Imports of goods and services	50.8	51.6	47.2	41.3	38.5	54.7	63.2	73.3	89.8	111.3	119.4[a]	130.0
	(18.1)	(17.6)	(15.1)	(12.8)	(11.4)	(15.0)	(15.5)	(16.9)	(18.5)	(21.3)	(20.5)	(20.5)

Gross domestic product (at market prices)	296.8	315.9	336.8	353.0	374.4	395.5	446.1	474.8	534.4	584.9	648.0	710.0
	(105.9)	(107.8)	(108.2)	(109.6)	(110.4)	(108.5)	(109.2)	(109.6)	(110.3)	(111.8)	(111.1)	(111.8)
Net factor payment abroad	16.5	22.9	25.4	30.9	35.3	30.9	37.5	41.5	49.7	61.7	65.0	75.0
	(5.9)	(7.8)	(8.2)	(9.6)	(10.4)	(8.5)	(9.2)	(9.6)	(10.3)	(11.8)	(11.1)	(11.8)
Gross national product	280.3	293.0	311.4	322.1	339.1	364.6	408.6	433.3	484.7	532.2	583.0	635.0
	(100.0)	(100.0)	(100.0)	(100.0)	(100.0)	(100.0)	(100.0)	(100.0)	(100.0)	(100.0)	(100.0)	(100.0)

SOURCE: Bank Markazi Iran, *National Income of Iran, 1959–1965* (Tehran, 1968), tables 38, 50, 54; idem, *National Income of Iran, 1962–67* (Tehran, 1969), tables 49, 65; idem, *Annual Report and Balance Sheet as [of] March 20, 1969*, table 18; and additional data provided by Bank Markazi.

NOTE: Details may not add up to totals because of rounding; data for various years are not always comparable.

a Estimated by the authors.
b Percentage shares in GNP are shown in parentheses.

in population. And although the target rate for the agricultural sector in the Fourth Plan period has been set at 4.4 percent a year, the actual performance in the first three years of the new plan has been merely 3 percent. The slow growth of agriculture—caused partly by unfavorable weather conditions—continues to be a major stumbling block against government efforts to raise the living standards of the rural population and to reduce the importation of food.

Employment has also increased in the past decade—at an average rate of approximately 2 percent.[4] Full employment has prevailed in the skilled and technical personnel market. As the labor force has grown at a somewhat faster rate, some pockets of unemployment have continued to exist, particularly among urban unskilled workers. But shortages have also prevailed in some highly skilled trades. To combat unemployment and underemployment, the government has diverted sizeable funds to education, manpower training, and to the upgrading of the labor force.

The high growth rate and the nearly full employment of skilled labor in recent years have been a direct result of substantial increases in both public and private investment. Annual gross domestic capital formation (exclusive of inventories) almost tripled in real terms between 1959 and 1970, rising from just over 17 percent of GNP to more than 21 percent (table 4.1). More than 60 percent of the investment increase has been in the public sector. As is clearly revealed by the data in this table, the share of private consumption in real terms declined from 74.6 percent to 60.2 percent of GNP during the 1959–70 period.[5] In contrast, the share of the government consumption expenditures, for both civil and defense purposes, has increased. Although civil expenditures have continued to claim a larger proportion of total government consumption expenditures, their relative share has of late begun to drop, with the result that the gap between civil and defense expenditures has somewhat diminished.

---

[4] According to a recent unpublished study, employment between 1964 and 1969 is estimated to have risen by 16 percent—implying a compound annual growth rate of about 3.9 percent. The same study also puts the overall unemployment rate in the first half of 1969 at 3 percent of the labor force—5.4 percent in urban centers and 1.5 percent in rural areas. This low rate has reportedly been unprecedented in contemporary Iran.

[5] It must be noted here that the 1959–61 data are not fully comparable with the data for later years. The above comparisons, rough as they might be, are nevertheless indicative of the underlying trends.

During the same period, the ratios of gross national savings to the gross domestic and national products have generally ranged well above 15 percent. This has in turn resulted in a relatively high rate of gross domestic capital formation.[6]

Gross national savings has accounted for a relatively high proportion of gross domestic capital formation (table 4.2). The ratio of GNS/GDCF demonstrates an initial upward trend beginning with 85.2 percent in 1959 and reaching a maximum of 123.6 percent in 1963, followed by a decline in subsequent years. This is somewhat matched by an inverted pattern of NFB/GDCF ratios (i.e., proportion of GDCF financed by foreign capital). The behavior of these ratios, particularly during the 1961–63 recession years, is influenced by the then prevailing slowdown in business activity.

Economic Stability and International Balance

The achievements in employment and income have been accompanied by a notable measure of price stability up until 1969–70. The cost-of-living and wholesale price indexes increased by about 2 percent or less a year between 1962 and 1969—an accomplishment uncommon for most countries during the same period (table 4.3). The stability was especially noteworthy over the 1965–68 period. In this period the wholesale price index remained virtually unchanged and the cost-of-living index rose by about 3 percent for the whole four-year period. The components of both indexes, too, showed no significant departure from the general average. In the wholesale index, only fuel prices rose substantially faster than the trend, whereas food, construction materials, and others followed the general trend more closely. In the cost-of-living index, medical care, housing, food, and house furnishings, rising faster than the general index, outpaced clothing, transportation, and others that rose by less

---

[6] Gross domestic fixed capital formation (GDCF) represents gross additions to the stock of fixed capital within the country, whether financed by the country's savings or by foreign funds. Given the relative importance of the former in augmenting the stock of capital, it can generally be regarded as the embodiment of national savings. It may be noted that the rate of capital formation is expected to reach a high of about 25 percent of GDP by the termination of the Fourth Plan. This has obvious implications for planning and growth. Assuming an incremental capital-output ratio of $2\frac{1}{2}:1$ and a population growth rate of 2.8 percent per annum, the gross domestic product per capita can be expected to increase by as much as 7 percent per annum in real terms.

Table 4.2. **Ratio Analysis of Capital Formation Data, 1959–70** (Percent)

Ratios	1959	1960	1961	1962	1963	1964	1965	1966	1967	1968	1969	1970 [a]
GNS/GDCF	85.2	89.2	103.0	122.2	123.6	108.4	104.8	100.4	97.5	85.2	96.4	100.0
GNS/GNP	14.6	16.6	18.5	17.6	18.4	18.1	19.7	19.5	21.3	19.8	20.7	21.2
NFB/GDCF	14.8	10.8	−3.0	−22.2	−23.6	−8.4	−4.8	−0.4	2.5	14.8	3.6	0

SOURCE: Computed on the basis of data in Bank Markazi Iran, *Annual Report and Balance Sheet as [of] March 20, 1969*, table 18, p. 44; *National Income of Iran, 1959–1965*, table 38, p. 88; and additional data provided by Bank Markazi.

NOTE: GNS = Gross national savings
    GDCF = Gross domestic (fixed) capital formation
    GNP = Gross national product
    NFB = Net foreign borrowing (net utilization of foreign loans and investments)
    GDCF = GNS + NFB
  [a] Estimated by the authors.

Table 4.3. Price Indexes, 1960–69 (Annual Averages; 1959 = 100)

Index	1960	1961	1962	1963	1964	1965	1966	1967	1968	1969
Cost of living	107.9	109.6	110.6	111.7	116.7	117.0	117.9	118.9	120.7	125.0
Wholesale price	102.0	102.2	103.6	104.0	109.6	110.6	110.0	110.2	110.9	114.7

SOURCE: *Bank Markazi Iran Bulletin* (November–December 1969), pp. 533, 553; and additional data provided by Bank Markazi.

than the overall level of prices. Some of the items in the food category with substantial price increases have been meat, poultry, and fish, whose index stood at 184.8 in 1969 compared with the index of food prices of 130.9 and the general index of 125 (1959 = 100).[7] Starting in 1969, however, price indexes began to show an upward trend, and the cost-of-living index rose by nearly 3.6 percent to 125 (table 4.3). A mild inflation of a similar magnitude continued during 1970 with no sign of an imminent slowdown.

One reason for this long period of price stability has been the substantial rise in imports financed partly by the country's growing foreign exchange earnings. Other reasons have been the existence of excess capacity and unutilized resources left by the 1961–63 recession; good agricultural crops between 1965 and 1968; expansion of new industrial facilities and output; and an appropriate monetary policy by the central bank. The stability of the price level has not, however, been matched by a balance in international payments. Despite a yearly increase of 18 percent in government receipts from oil between 1963 and 1969, and an annual increase of close to 11 percent in nonoil exports, the balance-of-payments position of the country has somewhat weakened. The main reason has been a perennial tendency for the growth of imports, conditioned by the high rate of domestic economic activity. In 1968 alone, for example, current foreign-exchange receipts increased by about 13 percent, whereas current foreign-exchange payments rose by about 30 percent (table 4.4). The 1969 situation showed no appreciable improvement, and in 1970 Iran was forced to make use of its regular and special drawing rights in the International Monetary Fund (in addition to obtaining substantial short-term credits and advances) in order to meet its growing foreign obligations.

The deterioration in the balance of payments is not in line with earlier projections of the planning authorities. This may be the logical outcome of a situation where the increase in the demand for imports (nearly 90 percent of which are made up of intermediate and capital goods) is bound to be greater than that of nonoil exports dominated by a few products (carpets, cotton, leather products, and mineral

---

[7] It should be noted that in the period before 1960, when the role of government in the management of the economy was considerably less than it is now, price-level fluctuations were more pronounced. For example, during the 1955–60 period the cost-of-living index rose about 40 percent, which was finally brought under control at the expense of slowing down the growth of the economy for a period of three years.

Table 4.4. Factors Affecting Iran's Foreign-Exchange Position, 1963–69 (Million Dollars)

	1963	1964	1965	1966	1967	1968	1969
Current receipts	618.8	701.4	817.3	940.8	1,175.5	1,325.1	1,518.7
Current payments	(548.1)	(759.2)	(932.3)	(1,089.2)	(1,387.7)	(1,802.8)	(2,070.7)
Capital account—net	(40.1)	(45.2)	19.2	119.7	214.1	382.2	397.3
Changes in reserves (increase)	(30.6)	(95.4)	52.2	24.4	3.1	99.7	70.9

SOURCE: Bank Markazi Iran, *Annual Report and Balance Sheet as [of] March 20, 1969*, pp. 112–13; and additional data provided by Bank Markazi.

NOTE: Changes in reserves include Iran's position in the General Account of the International Monetary Fund,
  Incidental payments by the oil companies (bonuses, payments resulting from changes in accounting procedures, etc.) are excluded from current and capital
accounts but included in the reserves account.

ores). However, with the emergence of new industries, especially the petrochemicals, metal products, aluminum, copper, and natural gas, it is expected that the composition of Iran's nonoil exports will undergo notable changes in the mid-1970s, thus helping to improve the external accounts.

A share of responsibility for the continued pressure on foreign exchange reserves should also be attributed to the essentially "self-sufficient" nature of Iran's development financing. During the 1959–68 period, the gross amount of annual capital inflow was about 2.5 percent of the GNP (as against 3 to 5 percent annual average for most developing countries). The scant reliance on outside help, thanks mostly to the sizeable oil income, is counterbalanced by the outflow of debt repayments and services which in some years (e.g., 1963–64) exceeded foreign loans and credits.[8] During the Third Plan period, the ratio of ex post private savings to GNP was, on the average, 11.3 percent a year. Given the average annual investment rate of 18.8 percent of the GNP in the same period, it can be seen that about 60 percent of total capital formation has been financed by private savings. As a result of increased investment activities in the Fourth Plan, however, the net inflow of external loans and credits for both development and defense purposes has begun to rise appreciably, and is expected to continue upward. The debt service, too, is expected to rise sharply over the next few years.

Individual Welfare

Because of the lack of sufficiently detailed data regarding the sources and composition of Iran's national income, the welfare and redistributive effects of recent development planning are not easy to assess, and are much harder to measure. Certain scattered indications, however, confirmed by further visual observations, seem to show some positive progress on both scores.

During the Third Plan period, the national level of living (as

---

[8] As is shown in table 4.1, the net average annual factor payment abroad during the 1962–69 period has been about 10 percent of the GDP. It is interesting to note at this juncture that during the same period the net inflow of foreign capital to Iran has accounted for less than 2 percent of the GDP. Although these two figures are not exactly comparable, they show that the Iranian economy as a whole has contributed much more to the national income of the "rest of the world" than it has received from other countries.

measured by private consumption plus access to free social welfare services) decidedly improved. According to official national income figures between 1959 and 1969, private consumption expenditures increased by 5.5 percent a year in real terms, or by about 2.7 percent per capita. During the same period government consumption expenditures on social and economic services rose by 10.8 percent a year, benefiting mostly the middle and lower socioeconomic strata of the population. Of these public expenditures on socioeconomic services, the outlays on education and health are particularly relevant. In the various development plans, special attention has been paid to education, both as a long-term investment in improved manpower and as an essential element in the people's higher level of living.

The Third Plan called for an increase in the number of children seven to twelve years old in primary schools from 1.5 million to a total of 2,225,000—roughly 60 percent of those of school age. At the end of the plan, this number actually reached 2.9 million. The success of the plan in exceeding its initial target was the result of stepped-up government effort, particularly in increasing the number of primary school teachers through the Literacy Corps. Secondary-school enrollment, targeted for 400,000 by the end of the plan, reached 579,000. Despite a population increase of about 2.8 percent per year during the plan, illiteracy was reduced to 70 percent from 75 percent. Among the basic achievements in the 1963–67 period was the extension of literacy and education to the remote areas of the country through the dispatch of literacy corpsmen.

In health, and particularly rural sanitation, the welfare impact of government activities also seems to have been noteworthy. According to official figures, during the Third Plan period, more than two hundred new rural clinics were established and the number of rural physicians increased from three hundred to one thousand, thus extending public health care at the village level to 50 percent of its potential as compared with a mere 25 percent at the outset. Environmental sanitation has improved through the Health Corps, and nationwide inoculations have been undertaken against smallpox, cholera, tuberculosis, and other public health hazards. A program of better nutrition has been established for about half a million primary school children.

With respect to increased fairness in income distribution (as a declared "secondary" objective of the Third Plan) certain progress —although slow and spotty—is in evidence. As a result of changes in the structure of the economy a new middle class has been fast rising, particularly in the urban centers. The land reform program, too, has unquestionably had some redistributive effects in favor of former sharecroppers and farm workers.[9] Private consumption in constant prices between 1962 and 1968 increased faster in the cities than in rural areas (6.9 percent versus 5.7 percent), but population also increased faster in the cities than in the country (4.9 percent versus 1.7 percent). Thus the gap between urban and rural per capita levels of consumption may conceivably have narrowed.[10] At the same time, owing to the relatively much lower rate of growth of agriculture as compared with all other sectors of the economy, the distribution has not been out of proportion with sectoral productivity. Also, to the extent that efforts have been made to increase the share of direct taxes in the ordinary budget, the burden of taxation on the low-income classes may be considered somewhat alleviated. According to the published budgetary accounts, direct taxes accounted for about 12 percent of the nondevelopment budget in 1963 and about 20 percent in 1970. And, more significantly, the rate of increase in direct taxation amounted to about 20 percent per annum, exceeding the annual growth rate of current nondevelopment expenditures.

As regards the increase in social security, the Iranian government seems to have reached new records. The urban, industrial workers, in addition to their rising real wages, have been given a share of the benefits of industrial production and profits. Civil servants, particularly those in contractual cadres without tenure, have been granted both increased salaries and generous pensions. Farmers and urban workers have been covered by some health and disability insurance. The government's Social Insurance Organization has extended its activities to cover, in addition to workers in the industrial sector, those who are engaged in service organizations. Thus the total number of persons insured by the SIO, which was 190,000 in 1957, reached 627,000 in 1970. Furthermore, the government

---

[9] See Bank Markazi Iran, *National Income of Iran, 1962–67*, pp. 127, 155–56.

[10] Since agricultural production has in some years increased much faster than rural population, and as more than 60 percent of the Iranians still live in largely nonmonetized and almost self-sufficient villages, a part of additional farm outputs has undoubtedly been consumed by the new farmer-producers. Yet inasmuch as the main part of annual economic growth has come from the industrial, oil, and services sectors, a substantial improvement in narrowing the income gap still requires further evidence.

insurance company (Bimeh Iran), with collaboration of the Central Organization for Rural Cooperatives, has put into effect a scheme to insure the villagers for up to Rls. 40,000 in the case of accident or death.

Insurance is also to be extended to all nonpermanent civil servants, who under the new Civil Service Employment Law cannot be granted official status. Under the provisions of the Fourth Plan the government will be able to control and coordinate private and public social welfare activities. Actions have been taken in regard to the expansion of retirement benefits to 400,000 government employees, welfare workers, and youth.

Medical expenses and other benefits paid to workers by the SIO increased from Rls. 1.2 billion in 1962 to Rls. 1.5 billion in 1967. Total payments for financial assistance and medical treatment during the period 1962–67 amounted to Rls. 7.2 billion. In 1969, according to the official figures, the Social Insurance Organization had eighty-seven hospitals, seventy-eight clinics, and eighty-six medical centers at its disposal. Similar social welfare activities were being carried out by such other organizations and semiprivate foundations as the Imperial Organization for Social Services in health care, and Red Lion and Sun Society in its new youth program.

Economic Diversification

Iran's vigorous economic expansion has been accompanied by significant structural shifts from a basically agricultural preoccupation to a diversified economy characterized by fast-moving industrial and services sectors. Concentrated at first on durable and nondurable consumer goods, the country's industrialization has been extended to intermediary products (e.g., steel, aluminum, petrochemicals), and even capital goods (e.g., motor vehicles, machinery, machine tools). Whereas much of the initial industrial expansion was oriented toward import substitution in the domestic market, recent additions have been bracing themselves for some, they hope, sizeable non-traditional exports. The planners' attention also has been directed of late to the expansion of industries based on local resources (e.g., fertilizers, chemicals, natural gas) in addition to those relying on foreign-based inputs (pharmaceuticals, tires, electrical appliances, synthetic textiles, cables). Both groups of industries are now moving ahead simultaneously. It is expected that the development of the latter group will soon lead to the growth of domestic raw-material industries like carbon black, nylon cords, and synthetic rubber.

Iran's concerted efforts toward economic diversification have been accompanied by marked shifts in the relative share of various sectors in total output and input. This may be taken to approximate the rate of the "assimilation" of the dynamic (oil) sector in the rest of the economy. For if the rate of assimilation is comparatively high, one would expect further "productivity-centers" or "growth-points" to emerge in other sectors of the economy. The expanding productivity centers will in turn influence resource allocation, thereby giving rise to structural shifts all over.

A high rate of assimilation, as exemplified by sectoral shifts, would also result in an increase in the flow of resources (oil and related products) from the oil sector to the indigenous sector. Since Iranian intermediate-goods industries require large volumes of energy or hydrocarbons as their raw materials, a greater assimilation of oil in the economy can be expected with reasonable certainty. We have already noted the impressive rate of increase in the domestic consumption of oil and related products. Here we shall attempt to present a brief sketch of recent inter- and intrasectoral changes in recent years.

Distribution of National Product among Four Major Sectors

The magnitude of shifts in sectoral output is portrayed in table 4.5. In this table, we have classified the economy into four major sectors: agriculture, industry, oil, and services. Agriculture includes farming, animal husbandry, forestry, fishing, and trapping. Industry incorporates mining, manufacturing, construction, water, and electricity. Oil includes the value added to the economy by the oil sector, including revenues accruing to the government and the oil companies. "Other sectors" embodies services, including transportation and communication.[11]

---

[11] It should be noted that this classification has been chosen primarily for the sake of convenience and consistency with published data and not because of its comprehensiveness. Kuznets, for example, defines industry to include transportation and communication. In other classificatory schemes, mining is combined with agriculture on the ground that it, too, is an extractive, primary operation. In still other classifications, the major components of

The major points of table 4.5 can be summed up as follows. First, the share of the agriculture sector in total product declined from 31.4 percent of the GDP in 1959 to 19.4 percent in 1969 and is expected to decline further to 17.2 percent by the conclusion of the Fourth Plan. Second, the share of industry proper, exclusive of transportation and communication, rose from 13.6 to 16.7 percent of GDP during the same period and is projected to reach about 20 percent by 1972. Third, the share of oil rose from 17.2 to 25.7 percent of GDP during the 1959–69 period and is expected to rise to over 31 percent by 1972. And fourth, the services sector showed no clear-cut trend during the intervening years, but registered a continuous rise in the last few years (1967–69); however, it is expected to fall to about 32 percent by the end of the Fourth Plan period as the oil sector (which does not require many local services) continues to increase rapidly.

The significance of these shifts among various sectors of the economy will become clearer when they are viewed in terms of sectoral shares in the total real product. A fall in the share of agriculture, for example, is the result of a rate of growth of agricultural output lower than that of aggregate output. Correspondingly, a rise in the share of industry implies a rate of growth of its output higher than that of aggregate output. These implications can be more readily grasped by computing the percentage growth of aggregate product (at factor cost) as well as the relative changes in the four sectors under consideration in constant prices. It will be seen that the relative change in agriculture, in real terms, is smaller than that of aggregate product, and the relative changes in industry and oil are invariably greater than that of aggregate product.

Both oil and "industry" have had a rate of growth appreciably higher than that of aggregate output. This differential in favor of these two sectors has contributed to enhancing their shares in gross domestic product. On the other hand, the rate of growth of agriculture during the period covered by table 4.5 has been less than that of aggregate output. This has resulted in a fall in the share of

---

services are dealt with separately, inasmuch as the different types of services are heterogeneous and cannot be legitimately lumped under one heading. We have ignored these distinctions and refinements, partly because of the unavailability of data and partly because they have no material bearing upon our main arguments. See Simon Kuznets, *Modern Economic Growth: Rate, Structure, and Spread* (New Haven: Yale University Press, 1968), pp. 86–93.

Table 4.5. Distribution of Gross Domestic Product among Four Major Sectors at Constant 1959 Prices, 1959–72 (Billion Rials)

Sectors	1959	1960	1961	1962	1963	1964	1965	1966	1967	1968	1969	1970[a]	1972 (projected)
Agriculture[b]	87.3	87.4	91.4	88.3	89.9	92.2	99.0	102.7	110.9	115.5	117.5	120.0	137.5
	(31.4)[c]	(29.5)	(28.8)	(26.4)	(25.4)	(24.6)	(23.6)	(23.1)	(22.2)	(21.2)	(19.4)	(18.2)	(17.2)
Industry[d]	37.9	47.7	47.7	45.1	51.2	57.4	65.5	72.5	84.9	93.5	101.0	113.0	154.2
	(13.6)	(16.1)	(15.0)	(13.5)	(14.5)	(15.3)	(15.6)	(16.3)	(17.0)	(17.2)	(16.7)	(17.1)	(19.3)
Oil (domestic value added)	47.7	52.9	59.5	65.5	73.8	75.6	87.6	98.7	122.9	132.2	155.5	180.0	250.5
	(17.2)	(17.8)	(18.8)	(19.6)	(20.9)	(20.2)	(20.8)	(22.2)	(24.6)	(24.3)	(25.7)	(27.3)	(31.3)
Other sectors[e]	105.3	108.3	118.8	135.1	138.7	149.6	167.9	170.8	180.5	202.7	231.8	247.0	257.3
	(37.8)	(36.6)	(37.4)	(40.5)	(39.2)	(39.9)	(40.0)	(38.4)	(36.2)	(37.3)	(38.2)	(37.4)	(32.2)
GDP (factor cost)	278.2	296.3	317.4	334.0	353.6	374.8	420.0	444.7	499.2	544.0	605.8	660.0	799.5
	(100.0)	(100.0)	(100.0)	(100.0)	(100.0)	(100.0)	(100.0)	(100.0)	(100.0)	(100.0)	(100.0)	(100.0)	(100.0)

SOURCE: Bank Markazi Iran, *National Income of Iran, 1959–1965*, table 29, p. 79; idem, *National Income of Iran, 1962–67*, table 16, p. 42; idem, *Annual Report and Balance Sheet as [of] March 20, 1969*, table 14, p. 39; and additional data provided by Bank Markazi. The 1959–61 series are not comparable with those of later years because of revision of data.
a Estimated by the authors.
b Agriculture includes farming, animal husbandry, forestry, fisheries, and trapping.
c Percentage shares in GDP are shown in parentheses.
d Industry includes mining and manufacturing, construction, water, and electricity.
e "Other sectors" includes transportation and communication, banking and insurance, wholesale and retail trade, ownership of dwellings, public and private services, and statistical discrepancies.

agriculture in gross domestic product. The rate of growth of services has only recently risen above that of aggregate output, implying a moderate rise in the share of this sector in total product. Our previous conclusions concerning the observed trends in the four sectors, therefore, seem to be confirmed.

Distribution of Labor Force among Four Major Sectors

In order to make a meaningful inquiry into the underlying shifts in the relative shares of different sectors, it is not enough to look only at changes in the output data. It is necessary to take into account the changes in the amount of productive resources employed in each of the main sectors considered above. In view of the fragmentary data available about different types of resources used in different Iranian sectors, we can concentrate on one factor only, namely, labor.

Table 4.6 summarizes the distribution of labor force among the four major sectors for the ten-year period from 1959 to 1968.[12] The data in this table reveal a distinct and clear trend in labor input away from agriculture and toward industry and services. The share of agriculture in the total labor force fell from 53.5 percent in 1959 to 45.9 percent in 1967, whereas the shares of industry and services rose from 20.6 and 24.9 percent to 27.7 and 25.8 percent respectively during the same period. The share of the oil sector declined some 0.4 percentage points from 1.0 to 0.6 percent.

The important point to be noted here is that although the *absolute* number of workers in industry and services has been on the rise, the absolute number of workers engaged in oil exhibits a clear downward trend. This is not surprising in view of the highly capital-intensive nature of the petroleum industry, and other peculiarities

---

[12] In considering the distribution of labor among the various sectors, it may seem advisable to consider the industry and oil sectors separately. The proportion of labor force employed in the oil sector is rather small compared with the total supply of labor and also in comparison with that portion of supply which is engaged in industry, inclusive of oil. This is so despite the relatively significant share of the oil sector in total product, thereby producing a rather high product per worker in the oil sector. In view of the fact that the oil industry is highly capital-intensive, the inclusion of the product-per-worker data in the oil sector with those of other sectors may distort the labor productivity data.

Table 4.6. Distribution of Labor Force among Four Major Sectors, 1959–68 (Thousand Persons)

Sectors	1959	1960	1961	1962	1963	1964	1965	1966	1967	1968
Agriculture	3,346	3,351	3,351	3,261	3,242	3,221	3,197	3,168	3,141	3,710
	(53.5)[a]	(52.5)	(52.5)	(51.6)	(50.5)	(49.4)	(48.2)	(47.1)	(45.9)	(53.1)
Industry	1,293	1,356	1,362	1,520	1,589	1,660	1,735	1,813	1,891	1,679
	(20.6)	(21.3)	(21.4)	(24.1)	(24.8)	(25.4)	(26.2)	(26.9)	(27.7)	(24.0)
Oil	61	58	52	46	43	43	44	43	42	41
	(1.0)	(0.9)	(0.8)	(0.7)	(0.7)	(0.7)	(0.7)	(0.7)	(0.6)	(0.6)
Services	1,555	1,614	1,614	1,494	1,546	1,597	1,652	1,706	1,762	1,560
	(24.9)	(25.3)	(25.3)	(23.6)	(24.0)	(24.5)	(24.9)	(25.3)	(25.8)	(22.3)
Total	6,255	6,379	6,379	6,321	6,420	6,521	6,628	6,730	6,836	6,990

SOURCE: *Bank Markazi Iran Bulletin* (Dey-Bahman, 1346 [1968]), p. 51, in Persian; Bank Markazi Iran, *National Income of Iran, 1962–67*, p. 58; and table 3.13.

NOTE: The 1959–61 employment figures have not been published in other sources. Accordingly, it was not possible to verify the accuracy of the data, which show no variations whatever in either total or sectoral (agriculture, oil-inclusive industry, and services) employment between 1960 and 1961.

Although it is not specified in the source, it is presumed that the 1959–61 industry data incorporate the labor force engaged in the oil industry, but not those engaged in local distribution and transportation. They have been adjusted to show the labor force engaged in industry (exclusive of oil) by finding the difference between the original oil-inclusive labor force and the employment figures pertaining to the oil industry.

The 1968 data are based on a recent unpublished study of the Iranian population and are not readily comparable with those of the preceding years.

[a] Percentage shares in total labor force are shown in parentheses.

of employment in the Iranian oil sector. Under these conditions the contribution of the oil sector to employment (and hence to a more equal income distribution owing to higher pay in oil activities) has been at best marginal.[13]

Productivity data in various sectors have been shown in table 4.7. As can be seen from this table, the ratio of product per worker in agriculture to countrywide labor productivity has steadily declined, whereas in industry this ratio has had an upward trend (following a decline during the initial recessionary years). The ratio for the services sector exhibits a decline during the second half of the period covered in the table, after some initial fluctuations. Since "industry" also covers construction (with many unskilled workers), handicrafts, and repair shops (whose product per worker usually rises slowly), the gains in "modern" industries must have been higher than the average figures indicate.

One point needs to be clarified here. Our empirical observations have been confined to changes in product per worker. These changes should not be equated with changes in productivity in general as if changes in product per worker were identical with changes in product per unit of total input. In general, when more than one factor input is involved in a given production function, the productivity (average and marginal) of any one factor can be measured only by holding the other factor(s) constant. However, in reality what we are likely to observe is a change in total product accompanied by changes in total factor inputs. This may not cause any difficulty as long as production function is linearly homogeneous (e.g., the widely used Cobb-Douglas function). Under these conditions,

---

[13] Strictly speaking, it is implausible to attempt to gauge the direct influences of the oil sector largely by measuring its impact on employment, given the capital-intensive nature of the industry and the technological constraints imposed on optimum capital-labor ratios. Under these conditions, the direct influences of the oil sector must ultimately lie in the creation and expansion of new markets from both demand and supply sides. To the extent that the oil sector provides a new and low-cost source of raw material (while the economy continues to be primarily dualistic) and purchases goods and services domestically (especially when the economy is capable of providing the capital-goods requirements of the dynamic sector), the magnitude of direct influences is augmented in a direct and important sense. Thus an important part of the direct influences of the oil sector is what may be termed "synergistic effects," a term sufficiently comprehensive to include all the immediate and induced effects of the oil sector. Thus the decline of labor engaged in oil is not only not alarming, but is clearly an insufficient and incomplete index of the magnitude of interaction between the two sectors via direct influences.

Table 4.7. **Ratio of Product per Worker in Major Sectors to Economywide Product per Worker, 1959–67** (Constant 1959 Rials)

Year	Economy-wide Product per Worker	Agriculture		Industry (Excluding Oil)		Oil		Services	
		Product per Worker	Ratio	Product per Worker	Ratio	Product per Worker	Ratio	Product per Worker	Ratio
1959	44,476	26,091	0.587	29,312	0.659	781,967	17.581	67,717	1.522
1960	46,449	26,082	0.562	35,177	0.757	912,069	19.635	67,100	1.444
1961	49,757	27,275	0.548	35,022	0.704	1,144,231	22.996	73,606	1.479
1962	52,839	27,077	0.512	29,671	0.562	1,423,913	26.948	90,428	1.711
1963	55,078	27,730	0.503	32,222	0.585	1,716,279	31.160	89,715	1.628
1964	57,476	28,625	0.498	34,578	0.602	1,758,139	30.589	93,675	1.629
1965	63,367	30,966	0.489	37,752	0.596	1,990,909	31.418	101,634	1.603
1966	66,077	32,418	0.491	39,989	0.605	2,295,348	34.737	100,117	1.515
1967	73,025	35,307	0.483	44,897	0.615	2,926,190	40.071	102,440	1.402

Source: Computed on the basis of data in tables 4.5 and 4.6.
Note: Ratios for 1968 were not computed because of the incompatibility of data.

if both inputs are increased by the factor t (t is any positive real number), output will also be increased by t, thereby leaving the productivities unchanged. Factor proportions must therefore remain relatively stable for our findings to be applicable to overall productivity. But there is little doubt that factor proportions in the various sectors of the Iranian economy have undergone changes in the recent past and are likely to continue to change, perhaps even at an accelerated rate. The available data on Iran are not sufficiently detailed to enable us to verify these ramifications. In agriculture, the stock of capital has obviously increased, while the labor force has fallen, but there is no doubt that the bulk of agriculture in Iran still remains highly labor-intensive. In industry and the services sector, the stock of capital has also increased, but so has labor— the labor freed from the oil and agricultural sectors as well as new entrants into the labor force. Thus both capital and labor have been on the rise without a clear indication if factor proportions have been shifted toward capital by a sufficiently large magnitude to support or invalidate the parallel between labor productivity and productivity of other factors.[14] One is tempted to believe that in the course of a decade or so covered by this study factor proportions are unlikely to be substantially disturbed, because such changes can only take place in the long run as new and different technological innovations are introduced. This, however, may not apply to Iran, where labor-saving investments have been made in a relatively short time.

Overall Sectoral Shifts

In short, there have been several important changes in the structure of the Iranian economy in recent years. The agricultural sector has shrunk while the industrial sector has expanded in terms of both

---

[14] In a massive pioneering study of the causes of growth (i.e., general, overall productivity) in nine countries Edward Denison has shown as many as twenty-three measurable factors to be associated with increases in output. The physical input of labor hours, the quality of labor training and skill, individual diligence, and shift of labor from low to high productivity employment are factors associated with labor. But there are many others attributable to capital, scale of production, technology, freer trade, and so on. Part of increased output is also related to such intangibles as better knowledge of production or management techniques. See *Why Growth Rates Differ: Postwar Experience in Nine Western Countries* (Washington: The Brookings Institution, 1967).

output and (labor) input. Moreover, the intersectoral shifts have been accompanied by major intrasectoral (i.e., product-per-worker) changes: productivity gains per worker have been higher in industry and lower in agriculture as compared with nationwide averages. These changes, brought about largely through the active participation of the government in development effort, are indicative of the intensity of recent innovative activity in the economy. They provide a fairly good approximation of the rate of innovation intensity in the indigenous sector. This is particularly significant when the structural shifts and product-per-worker increases are viewed against the time period during which they have occurred. Such shifts have historically come about after a much longer period of time.[15] The rate of innovative activity as reflected in productivity increases, and in shifts in the structure of the economy, seems to be significant by historical standards.

Development Drags and Growing Pains

Iran's impressive economic progress has been naturally, and expectedly, accompanied by a number of growth-induced problems to which a brief reference ought to be made at this juncture. First, owing to the essentially academic (as contrasted with vocational or technical) nature of Iranian high-school and college education, the emerging labor force has been caught somewhat unprepared for the challenge of rapid industrialization (including modern agricultural methods), and expansion of complicated technical services. As a result, unemployment or underemployment has remained as a vexing politicoeconomic problem mostly among high-school and some college graduates and unskilled workers, while there has continued to be an acute shortage of skilled manpower and managerial personnel in both the private and the public sectors.

Second, an already troublesome, and potentially alarming, migration of farmers and farm workers toward urban centers has begun. Within twenty years, for example, Tehran's population has increased from less than a million to more than three million

---

[15] In Great Britain, for example, the share of agricultural labor in the total labor force declined from 35 percent in 1801 to 23 percent in 1841; in France from 43 percent in 1866 to 30 percent in 1911; in Belgium from 24 percent in 1880 to 18 percent in 1910; in Japan from 52 percent in 1925 to 43 percent in 1942; and in the United States from 32 percent in 1910 to 12 percent in 1950. See Kuznets, *Modern Economic Growth*, pp. 106–7.

inhabitants. Higher wages and better working conditions in the cities, increasing redundancy of many farmers in the village, made apparent by mechanized agriculture, urban amenities (health care, educational and recreational facilities, etc.) not available in rural areas, and, ironically, the lower cost of certain staple items (e.g., bread, fuel, wearing apparel) in some major cities under public subsidies have been among the major causes of the rural exodus. As elsewhere in the rapidly urbanizing world, the onrush of villagers to the cities has caused many social problems—from a shortage of public utilities to slum housing, monumental traffic tie-ups, over-crowded public facilities, and urban unemployment.

Third, as a result of enlarged and improved education, increased communication, and more frequent urban contacts, the demand for welfare benefits has increased. Popular expectations for still larger and better social services (including public monetary and technical assistance to the private sector) have become rampant. And social frustrations have not been effectively abated. In the meantime, the country's success in improving health and sanitation, and in raising the levels of nutrition (at least for a significant majority) has resulted in a potentially bothersome population growth. Within one generation, the estimated annual growth rate of population has gone up to an alarmingly high rate of 2.8 percent and higher in the late 1960s, from a mere 1.7 percent in the early 1940s. Although the death rate is expected to fall further, there are no such indications for the birth rate.

Fourth, with the onset of the Fourth Plan, inflationary pressures (that had remained dormant for the entire Third Plan period) have begun to emerge once again. These pressures, which are generally a natural concomitant of a vigorous expansion, have, in Iran, been a response to several particular factors. Among them are a gradual elimination of industrial excess capacity created during the recession years 1961–63; the relatively high production costs in many infant industries, owing partly to high cost of material, high capital charges (including interest, depreciation charges, and "normal" profits), and high wages paid to skilled labor; unfavorable weather factors in some years and for some agricultural products; a continued high propensity to consume—particularly among urban masses; institutional and economic bottlenecks that generally accompany rapid growth; and the long gestation periods associated with some infra-structural projects (including roads and dams).

Finally, the tempo of development itself and its major direction toward modern industrialization have increased the demand for imports and put increasing pressure on the balance of payments. As was indicated before, merchandise imports have grown at an annual rate of more than 20 percent (or more than two and one-half times as fast as GNP), from slightly more than $547 million in 1962 to an estimated $1,500 million in 1969. The increased imports have been largely concentrated on raw materials, intermediate goods, and parts used in the production of durable and nondurable goods (e.g., synthetic textiles, tires, automobiles, electrical appliances, and some consumer items of short supply), partly to fight inflation. Although the oil sector, as the mainstay of exports, has increased substantially (almost twice as fast as GNP in the 1962–68 period) and nonoil exports, too, have had an annual increase of 11 percent, foreign payments have generally outrun exchange earnings. Fast import growth has also to some extent been the result of fairly liberal policies pursued by the government regarding import of goods and services as well as travel and foreign remittances.

The gap between current foreign exchange earnings and payments began to widen in 1968 because of increasing demand for further expenditures in both defense and development. Despite record foreign borrowing by both private and public sectors in that year (about $480 million by the government itself), a continued inroad has had to be made into the country's not-too-large foreign-exchange reserves.[16] Although foreign exchange reserves in 1969 increased by more than $60 million, there was an even greater increase in liabilities to foreigners, resulting in some deterioration in the country's overall reserve position.

Apart from the foregoing problems that have been more or less directly related to the development strategy, the country's defense posture (particularly its declared policy of keeping the Persian Gulf free from foreign intervention after the British withdrawal in 1971) has presented an exogenous constraint of significant magnitude. Defense expenditures have grown much more rapidly than GNP and have absorbed a very sizeable portion of current government

---

[16] As noted in the annual report of the International Monetary Fund for 1970, the rapid increases in aggregate demand resulting mostly from increased public and private investments, raising import levels and increasing current account deficits, forced the Iranian government in 1969 to cut back nondefense public expenditures, raise interest rates, and tighten bank reserve requirements.

revenues—although still trailing behind similar expenditures of most other countries in the region. A part of development expenditures, too (e.g., communications, housing, roads) although not directly defense-oriented, has been influenced by defense considerations. Defense requirements continue to hold, as in most other countries in the area, a key to further successes in the development field.

5

Good Luck and Good Design

As was outlined in chapter 2, the oil sector in Iran during the 1910–50 period failed to produce the type of responses called for in the Hirschman thesis. The unbalanced growth of the oil sector had no widespread repercussion on the economy either directly or indirectly. Whatever investment was made during this period was financed by meager domestic savings—forced and voluntary. The reasons are not difficult to find. As far as direct influences were concerned, the sudden and somewhat swift growth of the oil industry was simply beyond the adjustment capacity of the private sector. The indigenous economy was not directly aided much by the oil sector and consequently failed to respond fully to the rapidly growing imbalances created by that sector. Fiscal influences, also, were too small and slow to rise to make a significant impact on the public sector.

By contrast, the post-1954 experience seems to confirm the thesis that a major growth center may, under propitious conditions, provide the élan vital to make the rest of the economy grow and prosper. The Iranian development success in the last fifteen years appears to have been the result of several favorable factors: large and growing incomes from oil, intelligent political leadership, courageous social reforms, and a general development strategy reflecting national aspirations. It was through a combination of these factors that the large benefits generated by the main growth center, that is, oil, could be channeled to, and captured by, the entire economy.

Oil as a Resourceful Servant

By all accounts, the oil industry has been the backbone of Iranian economic development in recent years. In 1969, oil earnings by the Iranian government provided as much as 50 percent of all government revenues, and over two-thirds of current foreign exchange earnings. Foreign exchange received from oil activities in 1969 financed nearly 55 percent of total foreign payments (including loan repayments). Between 1963 and 1969, government revenues from the oil sector grew at a remarkably high average rate of 18 percent per annum—almost twice the rate of increase in GNP. In relation to the GNP, petroleum as a single industry is now second only to agriculture.

The contribution of the oil sector to gross capital formation by indirect influences is shown in table 5.1. It is clear from the data in that table that the share of oil revenues (indirect influences) earmarked for increased domestic investment has been on the rise.[1] The contribution of oil revenues to capital formation is expected to reach an unprecedented level of over 60 percent by the end of the Fourth Plan period. The basic role of the oil sector in the formation of capital is more forcefully brought out by the ratios in row 5 of table 5.1. The share of oil revenues allocated to the Plan Organization in total public investment outlays is significant indeed and constitutes the bulk of public investment expenditures.

Thus, Iran's economic development has benefited from the oil sector in more ways than one. Oil not only has served as a key element in raising both the magnitude and the tempo of capital formation, it has accomplished this striking task in a uniquely productive and "painless" fashion, matched by no other developing, nonoil economy. To be sure, the major portion of surplus resources

---

[1] The fall in the share of oil revenues going to the Plan Organization in the year 1962, which marked the trough of a three-year recession, is due to a decline in government revenues from other (nonoil) resources.

Table 5.1. Ratio of Oil Revenue Invested to GDCF and to Public Investment at Current Prices, 1959–69 (Billion Rials)

	1959	1960	1961	1962	1963	1964	1965	1966	1967	1968	1969[a]
(1) GDCF	48.1	56.5	55.2	47.7	50.3	62.4	81.1	89.7	115.0	131.7	156.6
(a) Private	31.9	40.5	37.4	33.0	34.3	43.5	47.8	54.2	60.5	69.9	77.5
(b) Public	16.2	16.0	17.8	14.7	16.0	18.9	33.3	35.5	54.5	61.8	79.1
(2) Oil revenue	19.4	21.4	21.7	25.7	27.7	36.4	39.5[b]	47.4	54.0	61.8	76.4[c]
(a) Share of Plan Organization[d]	9.7	11.8	12.4	10.0	16.3	22.3	27.5	34.4	41.0	46.8	61.0
(3) Ratio of oil revenue to GDCF (2) ÷ (1)	0.403	0.379	0.393	0.539	0.551	0.583	0.487	0.528	0.469	0.469	0.488
(4) Ratio of oil revenue allotted to Plan Organization to GDCF (2a) ÷ (1)	0.202	0.209	0.225	0.210	0.324	0.357	0.339	0.384	0.356	0.355	0.389
(5) Ratio of oil revenue allotted to Plan Organization to total Public investment (2a) ÷ (1b)	0.599	0.738	0.697	0.680	1.019	1.180	0.826	0.969	0.752	0.757	0.771

SOURCE: Bank Markazi Iran, *National Income of Iran, 1959–1965*, table 50, p. 98; idem, *National Income of Iran, 1962–67*, table 61, p. 77; idem, *Annual Report and Balance Sheet as* [of] *March 20, 1969*, p. 52, and table 43, p. 93; and *Role of the Oil Industry in Iran's Economy*, undated booklet published by the National Iranian Oil Company.

[a] Current price data for 1969 were obtained by inflating the constant price figures.
[b] Excludes **Rls.** 10.5 billion oil bonus.
[c] Includes **Rls.** 6.3 billion advance payment by the Consortium.
[d] Excludes utilization of oil bonus.

destined for investment in all dual economies is expected to come from the dynamic sector. And inasmuch as the causes of under-development can be attributed to the inadequacies of national savings and capital formation,[2] the dynamic sector's surpluses serve as a crucial element in the developmental effort. A vital distinction, however, ought to be made between surpluses generated in a *domestically based* dynamic sector (like food processing) and those supplied mainly from an *export-oriented* industry (like petroleum).

A home-based, dynamic industry, through its backward and forward linkages, may indeed set the pace for the rest of the economy to emulate. But the growth of these industries constitutes both a burden on domestic resources and a burden on further domestic consumption. Resources channeled into the growth industries must be drawn away from other investment alternatives—sometimes at the cost of neglecting some urgently needed improvements in the existing level of living; for example, education, health, and housing. Some socially necessary projects, in other words, have to be slowed down, or domestic consumption has to be kept in check, in order to support the growing sectors. The reinvestment of the accrued surpluses, too, must be made at the expense of other pressing welfare needs.

No such "back-breaking" constraints exist when the center is foreign-financed and export-oriented—that is "exogenously" dynamic. Thus, large investments in the Iranian oil industry have been supplied mainly from the sector's own surpluses and at no immediate cost to past or present generations of Iranians. Were such investments to be made out of Iran's own domestic resources (as they are now being made in a small scale by the NIOC in its own exploration, extraction, and refining activities), some other project would have had to be postponed, curtailed, or altogether neglected.

The royalties from the oil companies to the Iranian government,

---

[2] The "conventional wisdom" about development attributes the causes of underdevelopment primarily to inadequacies of capital formation. This view, which is part of some formally sophisticated growth models produced by Western economists, has recently been contested by the advocates of a multi-dimensional approach to development. The latter argue that underdeveloped countries are confronted with multitudinous constraints, only one of which may be capital. Furthermore, it is argued that there is no a priori reason why capital should be the greatest inhibiting factor to development effort. Such specifically needed inputs as skills, organization, and entrepreneurship may indeed exert a greater constraining influence on the development process.

too, have in a sense provided an enormous supply of "painless" savings generated in the nonnational sector. As a result, there has been no need for domestic consumption to be curtailed or restricted. In fact, both consumption and capital stock have managed to grow together appreciably and somewhat effortlessly. Sizeable additions to the stock of national capital have thus taken place without the kind of belt-tightening and forced saving frequently suggested in the literature on economic development as the key to economic progress. The inflow of large incomes from an industry in which nonoil resources of the country have played only a very small role has made rapid economic development possible without frustrating the profit-and-utility maximizing decisions of individual consumers and entrepreneurs. The overriding question facing the Iranian planners, in other words, has not been how to squeeze further savings out of the relatively meager domestic (nonoil) product and already low domestic consumption but the main problem has involved the much easier task of allocating these "God-given" surpluses wisely and efficiently among alternative investment outlets. Increasing expenditures and investments by the government have, furthermore, generated additional incomes and savings in other sectors, fostering the growth of private enterprise.

Another exceptional contribution of oil to the Iranian development has been the provision of enormous foreign exchange liquidity. Oil revenues not only have alleviated the problem of insufficient domestic savings as a major constraining factor in growth, they have also performed the equally difficult task of easing balance-of-payments difficulties. The plight of developing countries is, as is commonly known, not limited to the paucity of domestic savings and lack of investable funds. It is further reflected in the necessity of converting domestic savings into hard currencies that can be used for importing needed capital and consumer goods. Even if a developing country should manage to obtain investment blood out of its turnip-like consumption, it may still not be able to carry out its development tasks unless domestic savings can be readily turned into exports and foreign-exchange earnings. And, as every planner knows, the task of increasing exports is often much more difficult than that of domestic development itself. Domestic savings may be generated by the planners' ingenuity, by sufficient incentives, or by force. But it takes far greater planning wisdom and far more international cooperation to increase exports. Thus, when the bulk of a country's savings is originally in the form of foreign exchange, the complicated and herculean task of convertibility is almost automatically solved.

Despite these obvious advantages and contributions, owing to the lack of sufficient quantitative data the total real impact of the oil industry on the Iranian economy, and particularly in the share of GNP growth, cannot be accurately ascertained. Only some general observations can be safely made. First, the bulk of oil revenues is invested in public projects which have a direct bearing on the rate of growth. As detailed in the previous chapters, the share of public investment in the total fixed capital formation during the Second Plan was about 53 percent; in the Third, around 43 percent; and in the Fourth, projected to be 55 percent. These sizeable portions of the aggregate investment were, in turn, financed largely by oil revenues. Of an estimated total domestic capital formation of $15.6 billion during the 1951–70 period, oil revenues accruing to the Plan Organization have accounted for roughly more than 30 percent (see table 5.1 for details in more recent years). On this basis, one could conclude that about one-third of total growth in the 1951–70 period has been related to oil income.

Second, private investment, which has accounted for about half of total domestic fixed capital formation, has itself occurred partly as a response to expanding demand for consumer goods—both public and private. Public consumption of goods and services has been partly financed by that portion of the oil revenues paid to the general Treasury instead of to the Plan Organization. Private consumption has also been paid for partly by the oil income accrued to government employees as wages and salaries. Roughly one-third of the oil income during 1951–70 has been earmarked for current public expenditures. With a relatively high national marginal propensity to consume, the multiplier effects of this increased consumption have been significant.

Third, possibly of equal influence to that of aggregate consumption have been the accelerator effects of expenditure from oil revenues. Given the high ratio of real savings and investment to GNP—18 percent for most of the period and above 20 percent in some years —the magnitude of domestic capital formation has been unusually high for a developing country. This high level of domestic investment has taken place in a relatively short period of time and in a largely nonindustrial environment short of externalities and established ancillary industries. Thus each million rials of new public investment, creating demand for new goods and services, has induced fairly substantial initial investment to produce them. That is, although the original induced demand for such goods and services might not have been very large (and in a highly advanced economy might have produced only marginal effects on production capacity), in the Iranian setting it has had an accelerating effect on the economy, as most new factories had been set up from scratch. This has been true, for example, of the heavy initial outlays in the Ahwaz Pipe Mills to make pipes for the gas pipeline. Public investments in roads have given rise to large original expenditures for automobile assemblies, tire manufacturing, and other car accessories, all of which have been established for the first time and from the ground up. Much private investment that has taken place during the period under discussion has thus been influenced, if not actually triggered, by public expenditures out of the oil income.

All together, the receipts from the petroleum industry not only have accounted for about 12 percent of the Iranian GNP in recent years,[3] they have supplied between 46 percent and 50 percent of total government revenue. The receipts from the oil sector (including rial purchases of the oil companies) have accounted for more than two-thirds of the country's total foreign-exchange earnings on current account. Nearly half of the total national savings, estimated to be more than 17 percent of GNP at current prices in 1967–68, has been provided by the oil sector in the form of convertible currencies. And, as mentioned before, this abundant source of foreign-exchange earnings has been supplied without backbreaking burdens on the working population or belt-tightening pressures on the consuming public.

The availability of substantial amounts of foreign exchange thus not only has enabled Iran to obtain its needed capital and raw-material imports without the excruciating necessity of increasing its nonoil exports, it has also provided the planners with a strong and effective weapon to combat domestic inflation. The cost of living in the country has been kept from rising, unlike the situation in most developing countries, by periodic imports of consumer goods in short supply (meat, poultry, dairy products, fruits, butter, etc.), and by

---

[3] Total value added of the oil sector to the GNP (royalties plus wages and other contributions to the economy) has been estimated to be about one-sixth of the GNP during 1963–67 period. See Bank Markazi Iran, *National Income of Iran, 1962–67*, p. 38.

increasing imports of raw materials and intermediate goods to feed the flourishing consumer-goods industries—all financed by a relatively large supply of foreign exchange earned from oil. It is no coincidence that the accelerated rise in the cost of living starting in 1968 was preceded by new difficulties in the balance-of-payments situation.

In short, an unequaled share of responsibility for Iran's success in achieving a rapid rate of economic growth in the 1960s belongs to the oil industry. Without revenues from oil, the magnitude of public savings would have been much smaller, balance-of-payments difficulties painfully greater, inflation appreciably more rampant, and the standard of living not nearly as high. The appendix to this chapter examines Iran's share in the world petroleum industry.

Intelligent Leadership

The second, and perhaps equally crucial, reason for the success of Iran's development efforts during the 1960s has been its astute and courageous political leadership. Under the far-sighted and pragmatic direction of the shah, Iran has consistently moved several steps ahead of the type of politicoeconomic crises that have plagued other developing nations. As a result of this wise stewardship, Iran's political stability in the 1954–70 period was matched by only a few developing countries in the world—almost none in the Middle East.

The kingpin of the shah's successful one-upmanship, and the moving force behind his enlightened guidance, seems to have been his lifelong desire to reign (and as his critics charge, to rule) over a more progressive and more prosperous nation. And the most positive manifestation of this aspiration and vision has been his *White Revolution*, a peaceful program of radical socioeconomic reforms, initiated and supported by him since 1962, to raise the level of literacy, health, individual initiative, and economic well-being of his people.

These reforms, which are now widely referred to as the twelve-point "Revolution of the Shah and the People," originally encompassed six areas: land reform, forest nationalization, sale of government factories to the private sector, workers' participation in company profits, reform of the electoral law, and the establishment of a Literacy Corps. Two more corps—in health and rural reconstruction—and a House of Equity to handle interminable minor village litigations—were later added to the original six points. In 1967 the revolutionary program was extended to the nationalization of water resources, urban and rural reconstruction and development, and the reform of the country's educational and administrative systems.[4]

Some of these reforms (particularly the land reform program) contain significant implications for the purpose of our study and will be discussed presently. These programs, viewed as a whole, will have far-reaching repercussions that will touch upon virtually every major phase of the Iranian sociopolitical system in the years to come.[5] Their full and real effects, therefore, can be assessed only over an extended period of time.

Land Reform

The need for altering the land tenure system[6] in Iran as a fundamental condition for modernizing the nation's socioeconomic structure was not clearly recognized until Reza Shah assumed power in 1925. The growing centralization of economic and social affairs which characterized the reign of Reza Shah, together with a uniform land tax in lieu of the former numerous and diverse methods of assessing and collecting taxes, did much to undermine the power of the landlord and to provide the peasants with badly needed security from tribal raids. Beginning in 1927, some state lands were sold in Khuzestan and Sistan in an attempt to encourage and expand peasant proprietorship. However, for a variety of reasons these early experiments in land reform did not succeed in dislodging the large landowners from their traditional loci of influence. In 1939 a law was passed to modify the existing system of land tenure, but the precarious international situation prevailing at that time led the officials to decide against its immediate implementation.

---

[4] For the details of the last three programs see *The Revolution: New Dimensions* (London: Transorient, 1968). For the others see below.

[5] Most of the materials presented here have been taken from the official publications: *Panorama Iran* and *Iran Review*, October 1969, published by Embassy of Iran, Washington, D.C. It is interesting to note that the institutional reforms introduced in Iran are what Griffin and others have called the essence of economic development without which meaningful economic advancement may prove almost impossible. See Griffin, *Underdevelopment in Spanish America*, pp. 48–50 and 264–81.

[6] This section draws largely on A. K. S. Lambton, *The Persian Land Reform: 1962–1966* (London: Oxford University Press, 1969); and *Land to the People* and *Financing Land Reform* (London: Transorient, 1967).

The next important step in land reform came in 1951, when an imperial decree was issued for the sale of the royal estates to the peasants. The size of the lots to be distributed differed with differences in local conditions, fertility of the land, distance to urban centers, and so on. But the idea of land reform as a prerequisite for a meaningful political reform began to gather momentum after the oil settlement, and despite opposition from some landlords and a few religious groups, it eventually culminated in a bill in 1960. The new law limited the size of individual holdings to a maximum of 400 hectares (988 acres) of irrigated land or 800 hectares (1,976 acres) of unirrigated land. The implementation of the law, however, did not progress much beyond fragmentary actions, largely owing to the absence of sufficient data and administrative machinery for carrying on such a comprehensive land-distribution program.

The really serious step in land reform came on 9 January 1962, when the 1960 law was amended by a *decré loi* of the Council of Ministers (in the absence of Parliament) to (1) limit the size of individual holdings to one village, (2) fix the price due to the landowner for extra holdings to be sold to the peasants on the basis of previous years' taxes paid by him, (3) allocate the holdings among peasant cultivators without changing the existing field layout of the village, and (4) require membership in a cooperative society as a condition of the peasants' receiving the land. This law, together with the Supplemental Articles of 17 January 1963 (and the various other regulations related thereto), constitutes the legal framework for executing the land reform program. In Professor Lambton's view, the obvious pragmatism surrounding the law has been a unique feature of Iran's effort to undertake a gigantic step in reforming the nation's land tenure system.

The 1962 Land Reform Law was designed to bring about a major change in the landlord-peasant relationship, characterized by crop-sharing arrangements with a minimum of change in the environment and physical setting of production. The execution of the law came to be known as the first stage of land reform. Under this stage, a total of 16,000 villages representing about 19.5 percent of the arable land were purchased by the government from the landowners during the 1962–69 period and transferred to some 743,406 farm families (table 5.2). Although the number of villages directly affected by the first stage of the land reform constituted only a small proportion of the total number of villages in the country, the overall impact

Table 5.2. Results of the First Stage of the Land Reform Program, 1962–69

	1962	1963	1964	1965	1966	1967	1968	1969	Total
Number of villages purchased	3,705	5,002	1,605	2,991	1,571	136	451	539	16,000
Number of farm families receiving land	130,018	173,171	45,702	164,084	88,527	15,537	36,673	89,694	743,406
Installment payments to landowners (million rials)	339	223	426	1,305	547	73	64	60	3,037
Value of lands purchased (million rials)	3,339	—	—	—	—	231	317	292	9,868 [a]

SOURCE: Bank Markazi Iran, *Annual Report and Balance Sheet as [of] March 20, 1970*, table 77, p. 134, Persian edition.
[a] Details do not add up to total because of the unavailability of data for 1963–66.

117

of the reform was by no means small. The large landholdings of all major landlords were thus liquidated, with profound sociopolitical implications for the entire country.

Even though such vast and radical reforms in other parts of the world had generally been accompanied by strong resistance on the part of powerful landed interests (and there were apprehensions by many alarmists regarding a fall in agricultural output and a reluctance to push vigorously with industrial investment), the Iranian results surprised even the most optimistic. In many parts of the country, there was an expansion of mechanized farming and a rise of output due to better standards of cultivation. The tailor-made strategy to minimize the dislocations of the reform paid off handsomely.

Under the "Second Stage" of the land reform, landowners were offered a choice of five methods of settlement: (1) tenancy; (2) sale to peasants; (3) division of land on the same proportion as the crop-sharing agreement; (4) formation of agricultural cooperatives; and (5) sale of peasants' rights to landowners. Moreover, maximum individual holdings were set at 150 hectares (370 acres) and the reform was extended to cover the religious endowment lands.

Although there were many local differences in the execution of the "Second Stage," the peasants were given tenure, and their conditions were improved. They did not all receive ownership of the land (table 5.3), and the conditions under which the land was transferred to them were perhaps less favorable than those under the first stage.[7] Nevertheless, the "Second Stage" introduced an important period in Iran's agricultural history and rural cooperatives.

To create the degree of independence and self-reliance necessary for the peasantry to move out of the landlords' sphere of influence, a cooperative movement was clearly fundamental. This has been a gigantic task, especially in view of numerous cultural, social, manpower, and organizational difficulties. Despite these and other limitations, the number and effectiveness of rural cooperatives have increased noticeably in recent years (table 5.4). This has in turn provided an effective start in alleviating the problem of short-term credit. More than 1,800 high-school graduates have to date been trained to manage some 8,000 cooperatives covering 25,000 villages.

---

[7] Some 1.3 million rural households who acquired land under a thirty-year leasehold during the second stage have been given the option to purchase their own lands.

Table 5.3. Results of the Second Stage of the Land Reform Program, 1965–69

	1965	1966	1967	1968	1969	Total
Number of publicly-endowed lands leased to farmers	10,227	880	−1,902[a]	230	70	9,505
Number of privately-endowed lands leased to farmers	973	32	−76[a]	39	5	973
Number of small landowners who have leased their lands to farmers	129,648	73,203	3,990	1,861	3,120	211,822
Number of small landowners who have sold their lands to farmers	2,405	820	198	−9[a]	81	3,495
Number of villages in which land reform is completed	43,513	9,339	738	584	306	54,480
Number of farms in which land reform is completed	13,013	4,705	1,118	543	178	19,557

SOURCE: Bank Markazi Iran, *Annual Report and Balance Sheet as [of] March 20, 1970*, table 78, p. 135, Persian edition.
[a] Minus sign denotes correction of previous years' data.

Table 5.4. Rural Cooperatives Registered, 1963–69

	1963	1964	1965	1966	1967	1968	1969	Total[a]
Number of rural cooperatives	1,349	1,124	1,672	1,515	1,203	152	−286[b]	8,102
Number of members	138,196	102,808	118,836	172,442	151,082	173,134	139,342	1,399,762
Capital (million rials)	119	143	161	258	339	369	345	1,984
Number of loans granted by rural cooperatives to members	151,385	328,993	391,199	558,751	673,062	738,500	843,909	3,685,799
Amount of loans granted by rural cooperatives to members (million rials)	504	1,434	1,883	3,024	4,077	5,041	5,753	21,716
Number of rural cooperative unions	2	15	18	29	13	17	14	112
Number of members in rural cooperative unions	187	1,026	970	1,169	2,265	1,385	172	7,542
Capital of rural cooperative unions (million rials)	6	31	23	29	176	246	255	781

SOURCE: Bank Markazi Iran, *Annual Report and Balance Sheet as [of] March 20, 1970*, table 79, p. 136, Persian edition.
a Except for data on loans, totals include results achieved before 1963.
b Minus sign denotes correction of previous years' data.

In January 1966, the "Third Stage" of land reform was announced. The major aims of the new stage were (1) the expansion of agricultural output required for the industrial development of the country, (2) a rise in the per capita output and standards of living of the peasantry, and (3) the stabilization of food prices by improved marketing and production techniques. With the redistribution of land more or less complete, attention was thus being diverted to the full utilization of the nation's agricultural potential. Accordingly, water resources were nationalized in October 1967 and the three new ministries were set up to accomplish the goals of the "Third Stage": Ministry of Natural Resources, Ministry of Agricultural Products and Consumer Goods, and Ministry of Land Reform and Rural Cooperation.[8]

Furthermore, in an effort to overcome some of the problems arising from excessive fragmentation of landholdings (as a result of individual small ownership) the government has embarked on an ambitious agribusiness program. Under this program several large agricultural joint-stock companies have been organized in cooperation with foreign investors, and some agricultural corporations have been formed. Iran has been able to attract some foreign private investment from Europe and the United States for this purpose.

Gauged against the injustices, hardships, and inefficiencies of prereform days, the efforts of the government to bring about fundamental changes in the land tenure system have thus far been notable. A definite and irreversible shift has occurred in the landlord-peasant relationship; and, as a result, the immense sociopolitical power of the former absentee landowners has been broken. A large scale cooperative movement has been planned and partly set in motion. And, above all, the land reform has been carried out in an atmosphere of political stability and without major economic dislocations.

Profit Sharing in Industry

The welfare of industrial working men and women has been aided by the provision of a share of profits to be paid to those who work in the factories. The intent of the Labor Profit Sharing Law was to

---

[8] The latter was an amalgamation of the former Land Reform Organization of the Ministry of Agriculture and the Central Organization for Rural Cooperation.

grant wage increases based on higher productivity. The law has improved labor-management relations, and at the same time has given workers a better incentive to work more productively. Employers and employees are now charged by the new law to engage in collective bargaining in matters of wage rates, cost reduction, and increases in production.

The profit-sharing scheme is, in effect, a three-pronged weapon devised to increase the earnings, and hence the productivity of industrial workers. As a first choice a share of up to 20 percent of the net profit is to be distributed among workers in each factory. The other two alternatives prescribe extra compensations based on production norms, through higher productivity or less waste. In practice, efficient factories (mostly large, modern enterprises) have set both the salary and production norms, and served as a model for collective bargaining agreements in smaller, more traditional workshops.

Again, contrary to some dire predictions, the implementation of the law did not adversely affect investment in the industrial sector, and owing partly to improved labor relations, such investments have actually increased. Furthermore, the new incentive system has led to greater efficiency in production and helped develop collective

Table 5.5. Participants in Profit-sharing Plan

Area	Number of Plants	Number of Workers
Tehran	809	62,914
Isfahan	99	28,802
Mazandaran	74	10,543
Khorassan	87	6,884
Azarbaijan	80	4,748
Guilan	103	3,837
Fars	21	2,749
Khuzistan	65	2,211
Luristan	16	1,026
Kirmanshahan	28	930
Kirman	13	691
Hamedan	6	289
Sistan and Baluchistan	6	35
Kurdistan	5	33

SOURCE: Mohsen Shamlou, *Effects of the White Revolution on the Economic Development of Iran*, Scientific and Economic Publication no. 2 (Tehran: Tehran Chamber of Commerce, 1969), pp. 91–92.

bargaining experience. Table 5.5 shows the number of workers participating in the profit-sharing plan as well as the number of factories using the plan. [9]

Literacy Corps, Health Corps, and Development and Extension Corps

Illiteracy in Iran in the early 1960s was estimated at 80 percent. The number of children of school age was, in the meantime, increasing faster than the school system could accommodate them. According to a government estimate, primary school children were increasing by 10 percent a year, secondary school pupils by 14 percent in recent years. At the same time there were thousands of unemployed and underemployed high-school graduates in the cities. Here again an ingenious solution has been found to combat both of these social ills. Under a new educational scheme, these graduates have been assigned to the Literacy Corps to teach in the villages in lieu of a part of their normal military service. Any village ready to build a school through its own self-help could receive a literacy corpsman as a teacher. As a result, village schools have sprung up throughout the country, and thousands of young men were (after some teacher training) dispatched as elementary school teachers and multipurpose village-level workers to work in the villages under elected village councils.

Encouraging results have been obtained. By 1969 more than 50,000 young men (joined later by young women) had gone into the villages as teachers of "functional" literacy. Over 12,000 new schools had been built in which more than a million rural children were receiving elementary education, and more than half a million adults were attending basic literacy courses. According to an official estimate, in the 1970–71 school year some 15 percent of Iran's total population (4.5 million) were attending school. In cities, 90 percent of the children of school age were enrolled in primary schools; in rural areas, nearly 50 percent. The number of students in higher education, which totaled 58,000 in 1968 and 68,000 in 1969, now stands at about 80,000, exceeding the original target of the Fourth Plan by a significant margin.

---

[9] For further details see *The New Industrial Revolution* (London: Transorient, 1967).

The achievements of the Literacy Corps have led to the creation of the Health Corps. As the rural areas lacked teachers, so there was an even greater shortage of health aides. Under the new health scheme, the country was divided into regions, and several hundred mobile health teams were sent out, each headed by a qualified physician. A small base hospital was built at each regional center, around which a series of satellite mobile teams and clinics were set up. Once a health system was established in a region, the Ministry of Health could move in permanent physicians to carry out routine duties.

In practice, the Health Corps treats the sick, but is also concerned with the general health environment, public sanitation, preventive medicine, and modern hygienic practices. Health corpsmen reportedly tend the water supply, correct the sewerage, improve the bathhouse and the slaughterhouse, give basic inoculations, train midwives, and disseminate birth-control information and audiovisual material on health education. The Health Corps has thus served as a pioneer health service at the frontier of a new national health network.

For the battle against other rural problems of poverty and low productivity, the Development Corps has been organized. All available young agricultural school graduates are drafted into this corps in lieu of most of their normal national military service. Added to these draftees are some young engineers and architects. The Development Corps works on rural reconstruction as a supplement to the agricultural extension service. It helps the farmer help himself by showing the way to better agricultural practice. Veterinarians, engineers, and architects are all involved with improving the quality of village life. In 1969, over one thousand Development Corps men and women were recruited for a two-year tour of duty. So far more than *six* thousand Iranian women and many more Iranian men have participated in various corps. As a result, rural Iran has probably changed more in the past six years than it had for decades.

The result of these socioeconomic reforms has been a period of relative politicoeconomic stability conducive to economic progress and social rehabilitation. These reforms have so far offered a majority of the population tangible prospects for a better and more productive life. And this majority seem to have welcomed these prospects and responded to the challenges and opportunities involved. Thanks to these foresighted measures—instituted from the top and preceding public clamor—Iran has been able to introduce an impressive measure of dynamism in the previously dormant sectors of its economy while avoiding some of the disruptive (and sometimes destructive) consequences of dynamic growth.[10]

A Custom-made Strategy

The third crucial factor in Iran's development success has been a basically workable development strategy, which, in its totality, has largely offset the deficiencies of some individual projects and programs. Iran's efforts to develop economically have been primarily designed to diffuse the benefits of its major growth center (i.e., oil) throughout the economy. The Iranian development planning has aimed at increasing the direct and indirect influences of the oil industry on other indigenous sectors. This strategy has been partly aimed at providing some of the missing structural elements in the economy while strengthening the economy's other promising sectors. In this sense it may be said that the Iranian development plans have thus far been concerned with the first phases of attaining self-sustained growth through public investment financed by the oil income (indirect influences). The establishment of such new projects as petrochemicals, natural gas, and fertilizers that have a direct and relatively high linkage with oil may be taken as an indication that the planners are approaching a more advanced stage of self-generating development.

In more specific terms, the Iranian development strategy, as set forth in the third and the fourth development plans, has aimed at achieving a fairly high rate of increase in annual real income by building and strengthening a new, modern industrial and urban sector.[11] The growth targets have been a sociopolitical phenomenon around which the whole quantitative design of the plans has revolved. Since the overall growth rate has been exogenously fixed, it has determined other variables relevant to growth, including resources that must be forthcoming from the foreign-financed growth center.

---

[10] For further details of these programs see *The Literacy Corps* and *The Health and Development Corps* (London: Transorient, 1967).

[11] See Plan Organization, *Fourth National Development Plan, 1968–1972*. See also Baldwin, *Planning and Development in Iran*, chaps. 3 and 8, especially pp. 178–79; and particularly J. Price Gittinger, *Planning for Agricultural Development: The Iranian Experience* (Washington: National Planning Association, 1965), especially appendix 2: Iranian Agricultural Development Objectives.

With the increasing desire on the part of planners to maintain this high rate of growth, the magnitude of these foreign resources, too, has been pushed upward both by working for increases in the volume of oil production and through direct intervention and pressure for higher incomes.[12]

Although the high growth rate has been pivotal in the basic quantitative framework, a somewhat "balanced" approach to the underlying development design has also been a fundamental purpose of the development strategy. Although the planners' aim has not been to equate this balance with an even distribution of development resources among all sectors of the economy—at least not in the short run—their long-range perspective has envisaged such a balance.[13] The comprehensive and multifaceted goals of the recent Iranian plans further attest to the basically balanced nature of their strategy. The Fourth Plan, in particular, aims not only at accelerating the rate of economic advancement in all major sectors of the economy (e.g., chemicals, engineering products, and basic metals and minerals, including oil, water and power, and natural gas); it also includes substantial allocations for more and better public welfare services, greater diversification of the domestic economy, management training, and administrative reforms. Industry and industrialization, however, seem to have enjoyed greater administrative priority and popular support than warranted by considerations of comparative advantage.

A major aspect of this essentially balanced strategy has apparently been a conscious effort to spread industrial centers over a vaster area of the country—and particularly away from Tehran, where almost all major industrial and service establishments had tended to cluster during the Second Development Plan. The locational policy of

---

[12] The continued attempts and successes of the Iranian government at securing increased revenues from the oil companies may be said to have been prompted by the increasing needs of the government to accommodate the relatively high growth rate stipulated in the Fourth Development Plan.

[13] Indeed, the Third Plan briefly refers to an "optimum" rate of growth as that rate which "embodies certain balances between successive development plans as well as within each plan. In other words, an optimum growth concept provides a better perspective in judging programmes such as manpower training, agrarian reform, health, and generally investment in overhead and infrastructure. The response of such programmes may be slow or even negative in the initial years, but quite significant when a longer time-horizon is kept in view." Plan Organization, *Outline of the Third Plan, 1341–1346*, pp. 48–49.

decentralization has been followed by the government itself in the erection of its new industrial plants; it has also been effectively enforced in the issuance of licenses for private operations. Another part of this strategy has been a notable drive toward self-sufficiency in certain light consumer-goods industries. Still a third part of the basic strategy has been the establishment of an extensive industrial base in such heavy industries as steel, petrochemicals, aluminum, and machine building. Although some western economists continue to question the wisdom of this particular industrial policy in terms of economic efficiency, Iranian planners seem to find an overall justification for their approach.

Some concerted efforts, to be sure, have also been made to reconcile the competing claims of agriculture and industry on the available resources. The planners could not ignore the overlapping aspects of these two main strands of the development program, since they are interrelated administratively as well as conceptually. Yet they seem to have felt that since these two sectors do not always have the same constituencies, leaderships, protagonists, and preoccupations, their active promotion in a development program is consistent with both a balanced strategy and self-sufficiency in the international market. The planners seem to have felt that, in industry, the main bottlenecks that determine the rate of output expansion (via a diminishing marginal rate of productivity) are essentially specialized labor and capital equipment. In agriculture, on the other hand, the major obstacle to expansion has appeared to be mostly organizational and structural; that is, the whole spectrum of human relations governing agricultural production and marketing has been thought to be incompatible, and to a large extent inconsistent, with nonmarginal increases in agricultural output.[14] The solution of these problems, in the planners' opinion, has apparently called for essentially different talents and capabilities.[15] For one thing, they argue, the rectification of the agricultural constraints does not require heavy outlays of foreign exchange—at least not to the extent required

---

[14] For a similar analysis regarding the development programs of India, see Lewis, *Quiet Crisis in India*, pp. 45–49.

[15] It has to be admitted that the distinction drawn between agricultural and industrial input requirements, like many such distinctions, is not always clear-cut and absolute. A major weakness of such a distinction will appear when both agriculture and industry claim the time and the resources of high-level planning units. But this is apparently not significant in relation to total resources required by each sector alone.

by the industrial sector. For another, it does not call forth the many specialized human resources claimed by industry.[16]

In Iran, these two main sectors have thus both been stressed, although unequally. Since the capital-output ratio in the Iranian agriculture is usually higher than that in industry, the aim has been to provide an adequate supply of agricultural products and sufficient demand for surplus rural workers, whereas the major impetus to growth has been meant to come from industry and mining. The reason for this approach is easy to see. Since it has been estimated that output and income in the course of the Fourth Plan (1968–72) will increase at an average of 9 to 10 percent per annum, the country's demand for food is also expected to increase at an average rate of approximately 5.5 percent per annum. So Iran is expected to boost its food output by generally the same rate.[17] This rate of growth has to be achieved if the total expansion effort of the economy is not to slow down or be completely thwarted. This is indeed what the basic philosophy of balanced growth, stated in stronger and more positive terms, ultimately boils down to. The planners have felt that a widening gap between agricultural production and consumption is thus inconsistent with this fundamental fact, and may very well be at variance with other socioeconomic objectives.

Equally important in the planners' long-term strategy has been a sensible, and essentially nonideological, approach to the role of the state in the economy. Instead of getting bogged down in the interminable controversy of public versus private enterprise, the govern-

---

[16] For example, according to the official estimates, the number of agricultural engineers and technicians required in agriculture during the Fourth Plan period is slightly over 8,000, whereas the number of industrial engineers and technicians needed during the same period is almost twice that. The "skilled and specialized" industrial labor is listed at 171,000, and the number of "trained farmers" at more than 290,000.

[17] The percentage growth in aggregate demand for food was calculated on the basis of data published by the Bank Markazi Iran, Plan Organization, and the estimates of food consumption in Gittinger, *Planning for Agricultural Development*. The following equation was used:

$$\Delta C_f / C_f = (\Delta Y / Y)[(\Delta C_f / \Delta Y) \cdot (Y / C_f)]$$

where C_f = Consumption of food
Y = National Income
$[(\Delta C_f / \Delta Y) \cdot (Y / C_f)]$ = Income elasticity of demand for food. It is needless to add that in the above formulation we have neglected, as have the planners, the effects of both price- and cross-elasticities. See Baldwin, *Planning and Development in Iran*, p. 79. The target rate of agricultural growth as published in the Fourth Plan is 5 percent per annum.

ment seems to have found a workable compromise through the division of responsibility. In its policy of "growth with compassion," the government has monopolized certain economic activities (e.g., oil, forests, water, tobacco, and communications). It has shared the task of the development of others (e.g., steel, copper, cement, textiles) with private enterprise. It has volunteered to sell to private interests some of its own factories in already established industries, while taking the initiative in newer ventures. It has provided almost all of the major infrastructural facilities while leaving many profitable activities to the private sector.

The private sector's response to this challenge has signaled a notable break with tradition. At about the time of oil nationalization, Iran was still an underdeveloped country with prospects for only moderate growth—chiefly in the public sector. The industrial private sector was largely dormant. The few light (consumer) industries, developed between the two world wars, were chiefly in government hands and not all in very healthy condition. Private wealth and economic activity were based on landownership and the bazaar trade. As in all underdeveloped countries, industrial entrepreneurship was even less developed than private industrial enterprise. There was no real domestic private banking. The private sector was characterized by the so-called trader's mentality—that is, short-term private gains through foreign trade or land speculation.[18] For all these reasons, it was believed that the private sector—composed mainly of old-fashioned bazaar merchants—could not play any significant part in Iran's modernization and industrial development.

Yet as has turned out, a sizeable share of responsibility for Iran's economic growth since 1954 has been with private entrepreneurs. In the Third Plan, the private sector was expected to invest some Rls. 30.5 billion in industry and mining—almost 47 percent of total investment in that sector. Actually, however, private investment was more than one and one-half times the expected amount. Encouraged by private response to the government's call for increased investment, the planners expect more than Rls. 120 billion of industrial investment to be undertaken by the private sector during the Fourth Plan period.

There is no single conventional explanation for the sudden rise of private industrial entrepreneurs in Iran. The Iranian experience

---

[18] See *Iran* (New Haven: Human Relations Area Files, 1957), chap. 11.

seems to have cut across many hypotheses and matched the experiences of many different countries. Conventional hypotheses assume that (1) industrial entrepreneurs usually emerge from among educated, Westernized classes in a traditional society; (2) they are often not only economic innovators but recognized pioneers in other areas; (3) they develop from a once prestigious group that has become déclassé; or (4) they are often special minorities different from the society's traditional elite.[19] The Iranian experience lends credence to most of these hypotheses.

Iran's new entrepreneurial corps includes, first, a number of European- and American-trained young Iranians born mostly to old "merchant" families and frequently managing old family fortunes. Second, many among Iranian private entrepreneurs are successful professionals (doctors, lawyers, engineers) who indeed were pioneers in introducing modern science and technology to Iran. Third, managing successful private businesses for themselves and their families are many former landlords who have found new profits and new status in such ventures as mechanized farming or other private pursuits. Not a part of this class, but falling into the category of previously nonbusiness people finding "new life" in business, are many retired civil servants and former military officers who are using their administrative and technical knowledge and experience in new private ventures. And, finally, a number of Iran's successful entrepreneurs have arisen from the lowly civil-servant families (e.g., office boys, messengers, and clerks) that never enjoyed the status, prestige, or power of the Iranian elite—the landed gentry, the military, the clergy, or top-ranking political leaders. In short, the challenge of public planning and leadership has found singularly extensive response from many walks of life in the new Iranian society.

The enthusiastic reaction of the private sector to government planning in the past has been the basis of its continued role in the present and future plans. In the Fourth Plan, the private sector is encouraged to assume the lead in industrial investment. Public investment is restricted to the "key industries." The overall direction of new private investments is against imports and in favor of exports. Licensing of new industrial units is determined, among other things,

---

[19] For a discussion of these theories and their applications to Pakistan see Gustav F. Papanek, *Pakistan's Development: Social Goals and Private Incentives* (Cambridge: Harvard University Press, 1967), pp. 46–55.

by the domestic-value contribution of the industry, its efficiency, and its location. Thus the new industries are expected to have a domestic value added equal to at least 65 percent of the selling price of the finished product or should import new technology with favorable employment or other broad economic effects. Moreover, they are to be sufficiently competitive with foreign-based industries and in general to be located outside a radius of 120 kilometers from Tehran. To promote exports, a wide variety of measures are currently being contemplated by government officials, including tax exemption on incomes earned from exports, credit facilities with low interest, a reduction of up to 50 percent in port dues and charges, export credit insurance, remission of import duties on the products re-exported, standardization, and export bonuses. These policies, to remain in effect for at least the balance of the Fourth Plan period, are expected to provide a reasonably stable and calculable environment for the private sector to expand. The full impact of these policies still remains to be seen.

Short-run "Fine Tuning" and Long-term Approach

As was indicated in the previous chapter, Iran's economic progress has been accompanied by some obvious growing pains. Although the more deep-rooted among these difficulties continue to tax the planners' efforts and ingenuity, the government's response to many of these short-run snags has been both quick and decisive. To relieve the skilled-manpower and management shortages, specific training programs have been initiated by several ministries (e.g., Post, Telegraph, and Telephone) for their own needs. The Ministry of Education has revised high school curricula in order to step up vocational training at that level. The Ministry of Labor, in cooperation with private industry, has established a number of workers' skill-improvement schemes.

To combat inflation and balance-of-payments strains which developed at the end of the 1960s, a series of monetary and fiscal measures have been put into effect. Nondefense expenditures have been kept partly in check; attractive interest rates have been offered on government securities; credit has been manipulated by orthodox central banking operations; export drives have been intensified; and new markets for crude oil have been sought in Europe, Asia, and America.

To stabilize population growth, a far-sighted family planning program has been launched and gradually expanded. Rural migration has been dealt with by expanding social services (education, health, and recreation) at the village level through the various "revolutionary" corps; through the creation of centers or "poles" of development in rural areas; and through further land reform improvements. All in all, the Iranian planners seem to have shown an increasing awareness of these and other difficulties involved in the present development efforts.

On the whole, the basic Iranian development strategy has been successful in creating a modern, dynamic, mechanized sector whose benefits in both higher income and greater employment opportunities have permeated the traditional (largely agricultural) and slow-growing sector. The heavily import-dependent production of finished consumer goods has been maintained by means of tariff protection, public subsidies, and publicly sanctioned high retail prices designed to ensure profitability. The planners have hoped that once these new industries begin to mature and markets expand sufficiently, costs will decline and the need for these universal props will diminish. Although there is theoretical support for this optimism, the actual behavior of the Iranian cost factors has yet to bear this out. Also remaining to be seen is the extent to which the top planners can continue to maintain rapid economic growth with stability under the intensifying pressures of interminable new public projects, and in the face of the almost insatiable expectations of a demanding population.

As the need for "fine-tuning" becomes more strongly felt when the economy takes on added complications, so does the need for a better custom-made strategy become more apparent when typically Western solutions prove incapable of solving Iranian problems. Many of these problems have so far shown, and are now showing, a bewildering immunity to Western remedies. And the key to their solution—and indeed to the whole future of economic prosperity in Iran—lies in finding genuinely Iranian remedies suitable to the Iranian environment.

Planning Inadequacies

Our arguments in this chapter, and indeed throughout the book, should not be construed to mean that the Iranian planning machinery and process have been ideally flawless. In fact, as has been suggested before, there have been several not altogether unexpected weaknesses in all four plans. Some of these shortcomings are inherent in all simplified plans that do not make sufficient use of modern econometric techniques. Others are chiefly a reflection of the paucity of relevant and reliable data. Part of the problem is also due to a shortage of planning experience.

Judged by the more rigorous standards of comprehensiveness, consistency, and commitment, Iranian planning cannot be expected to be in the same league with the more sophisticated varieties. In terms of comprehensiveness the planning scope has been limited: all four plans have dealt mainly with the *public* sector. All major priorities, too, have been planners' priorities: all important planned projects have been public projects. The projections, expectations, and targets for the private sector have, at best, been "indicative" and expectedly unenforceable. Even within the framework of "indicative planning," the role of fiscal and monetary inducements in private decisions has not been made explicit. And uncertainties as to future government economic policies affecting private business —although substantially alleviated by repeated government assurances of policy continuity—have not been altogether eliminated. There has been very little formal consultation with private business leaders or organized private groups.[20]

From the standpoint of internal and external consistency, even the Fourth Plan (the most elaborate of all so far) has lacked a certain technical sophistication. Its input-output calculations, its projections of demand, and its estimates of supply bottlenecks have been rather rudimentary. The relationships between sectoral targets and policy variables have been neither causative nor conditional. Specific policy instruments to cope with emergencies have not been adequately described. And above all, the plan performance has been subject to little formal supervision and control. All four plans have shown extensive flexibility with regard to their sectoral targets and sectoral projects.

Commitment has been the third problem. The Iranian plans, including the Fourth, have had no firm commitments for implementation on the part either of the private sector (which has been mostly

---

[20] For similar difficulties of planning in free enterprise economies, see C. F. Diaz-Alejandro, "Planning the Foreign Sector in Latin America," *American Economic Review* (May 1970), pp. 169–79.

independent of the plans) or of the public agencies (which have been mainly dependent on plan outlays but largely free to propose different projects). Since the framework of the plans has been prepared by the Plan Organization, often with little or no direct participation from regular government ministries (and virtually none from the private sector), the responding agencies have felt little or no commitment to follow the planners' lead. The close relationships between the Plan Organization, Ministry of Finance, Bureau of the Budget, executive ministries, and public enterprises have also often hung on a delicate balance of personal friendship and cooperation among top leaders—and in spite of occasional interagency bickerings and rivalries.

Not only has the realization of the planned pattern of development been hampered by the lack of identification between the planners and the "responding forces," but the planners themselves have not been able to foreclose major unplanned projects. The land reform program, for example, was superimposed on the Third Plan in a totally exogenous—albeit inevitable—manner, with far-reaching consequences for both planned outlays and planned targets. The heavy new requirements of defending the Persian Gulf (caused by the planned British withdrawal from the area), and increased anti-Iranian activities on the part of some Arab neighbors during the first two years of the Fourth Plan, have introduced some unforeseen modifications. The possibility of oil revenues' falling short of the planned targets during the plan period has also played havoc with the coherence and attainability of the plan targets.[21] Sometimes, also, some new projects, exogenous to the original plan frame, have been given added priority under new circumstances or pressures. The Fifth Plan, now under preparation, has thus a good deal to learn from these past experiences.

To repeat, a large number of these inadequacies have been the result of statistical and information deficiencies common to most developing countries. Part of the difficulties in plan preparation and implementation also have reflected the novelty of the planning concepts for the Iranian civil servants. Their cultural aloofness to a planned way of life has been another factor. Yet in the context of the Iranian political economy this general flexibility has not been

---

[21] For the requirements of effective planning within the framework of free enterprise, see S. S. Cohen, "From Causation to Decision: Planning as Politics," *American Economic Review* (May 1970), pp. 180–85.

without merit. It has enabled the planners to maneuver within a total outlay, among projects whose priorities have been altered by unforeseen turns of events. It has also enabled experienced project managers and program administrators to correct original planning errors and omissions.

Appendix: Iran in the World Petroleum Industry

World production of petroleum has reached unprecedented levels, surpassing even the most optimistic projections of ten years ago. As the 1960s began, oil experts were projecting a jump of 10 to 13 million barrels a day in the production of crude oil by the close of the decade. That target was reached by the middle of the 1960s! World crude oil production of over 16 billion barrels in 1970 was more than double that of a decade earlier. The current forecast is that world demand for crude oil will exceed 83 million barrels per day by 1980, up from 45 million barrels per day at the end of 1970.[22]

Although output in virtually every producing country has expanded, the increase has not been the same everywhere. There have been considerable shifts in the relative share of different countries in aggregate world oil supply as well as in reserves. The share of the Western Hemisphere in total proved reserves has declined, whereas that of the Middle East and North Africa has registered substantial gain. At the end of 1970, the Western Hemisphere accounted for about 12 percent of total world proved oil reserves—down from about 20 percent in 1958 and 50 percent in 1948. On the other hand, the share of Middle Eastern countries jumped from 44 percent of the total in 1948 to about 56 percent in 1970. African proved reserves, which were almost nil in 1948, grew to about 1.5 percent in 1958 and slightly over 12 percent in 1970.

Iran currently accounts for about 11.4 percent (70 billion barrels) of total world proved reserves—a rate which has been fairly stable in the recent past. These reserves place Iran third in the world after Saudi Arabia (with 80 to about 140 billion barrels) and the Soviet Union (with about 77 billion barrels).[23] Even though the estimated

---

[22] *Oil and Gas Journal* (29 December 1969 & 28 December 1970).

[23] This marks a deterioration in the relative oil reserve position of Kuwait, which had more reserves than Saudi Arabia a decade ago. See *International Economic Review*, Monthly Bulletin of the First National Bank of Chicago, May 1969, p. 3.

proved reserves of Iraq rose steeply between 1948 and 1958, the current available data places Iraq (with 32 billion barrels) after Kuwait (about 67 billion barrels), the United States (37 billion barrels), and closely trailed by Libya (29 billion barrels).[24]

The production of crude by various countries has followed trends roughly similar to the growth in proved reserves. The fall in the share of the Western Hemisphere from 59 to 37 percent of total crude produced in the world in the decade of the sixties is in contrast to the gains registered by the Middle East and Africa.[25] The proportion of crude currently supplied by the latter regions is about 30 percent and 12 percent respectively—up from about 23 percent and a negligible output at the start of the decade. The Communist nations supplied about 15 percent of total crude in 1959, compared with their output of nearly 17 percent in 1970, with over four-fifths produced by the Soviet Union (see table 5.6). In 1970, Iran ranked first in the Middle East and fourth in the world by producing about 190 million tons, or about 8 percent of the world total,[26] thus regaining its leading position in the Middle East—a position which was lost during the oil crisis of 1951–53.

Petroleum exports as well as receipts by governments of oil-producing countries (tables 5.7, 5.8) have closely followed the tempo set by production. Consequently, Iran is currently beginning to emerge as the leading exporter of petroleum in the Middle East, with record annual earnings. Furthermore, it has gradually replaced Venezuela and the early indications now available for 1970 show that Iran has "firmly ousted Venezuela from its historic position as the world's biggest oil exporter."[27]

There have also been marked shifts in the distribution of refining capacity among the various countries. The shares of the Western

---

[24] The oil reserves of the Communist countries are estimated at about 100 billion barrels. Approximately 77 billion of these reserves are in the Soviet Union and most of the remainder is in mainland China. See *Oil and Gas Journal* (28 December 1970), p. 93. For the origin of the latest estimate for Saudi Arabia, see *Petroleum Intelligence Weekly* (30 June 1969), p. 3.

[25] In 1948 the Western Hemisphere accounted for over three-fourths of total crude oil produced in the world, whereas the Middle Eastern countries supplied only 12 percent—largely from Iran and Saudi Arabia. See *International Economic Review*, pp. 1–2.

[26] The first three producers in the world were, respectively, the United States (534 million tons), the Soviet Union (353 million tons), and Venezuela (193 million tons). *Petroleum Press Service* (January 1971), p. 7.

[27] *Kayhan*, international edition, 4 March 1970, p. 1.

Table 5.6. Estimated World Crude Oil Production, 1959–70
(Million Metric Tons)

Area	1959	1964	1968	1969	1970
Middle East	231	387	573	634	712
Iran	(46)	(84)	(142)	(169)	(190)
Saudi Arabia	(54)	(86)	(141)	(148)	(175)
Kuwait	(70)	(107)	(122)	(129)	(138)
Iraq	(42)	(61)	(74)	(75)	(76)
Others	(19)	(49)	(94)	(113)	(133)
North America	404	458	560	572	604
U.S.A.	(379)	(417)	(503)	(510)	(534)
Canada	(25)	(41)	(57)	(62)	(70)
Eastern Europe and China	147	250	338	363	393
USSR	(130)	(224)	(309)	(329)	(353)
Others	(17)	(26)	(29)	(34)	(40)
Africa	3	76	183	231	272
Libya	—	(42)	(124)	(150)	(159)
Others	(3)	(34)	(59)	(81)	(113)
Caribbean area	160	193	206	206	211
Venezuela	(147)	(177)	(187)	(187)	(193)
Others	(13)	(16)	(19)	(19)	(18)
Other Latin American countries	27	41	53	55	56
Far East	26	31	45	57	70
Indonesia	(18)	(23)	(29)	(40)	(45)
Others	(8)	(8)	(16)	(17)	(25)
Western Europe	13	18	17	17	16
World Total	1,011	1,453	1,976	2,134	2,334

SOURCE: *Petroleum Press Service*, (January 1970 and January 1971).
NOTE: Details may not add up to totals owing to rounding.

Hemisphere (including the United States) and the Middle East (including Iran) in total world refining capacity have declined, while Europe and Japan have gained in world importance. Annual crude refining capacity in the non-Communist world now exceeds 16 billion barrels—up from 7.3 billion in 1958 and over 14 billion in 1968. The United States now accounts for about 30 percent of the total, whereas Europe and Japan together account for upward of 40 percent—a nearly complete reversal of positions about a decade ago. Iran had over 5 percent of the non-Communist world refining capacity in 1948, over 2 percent in 1959, and only 1.4 percent in 1970. About one-fifth of the refining capacity in the Middle East,

Table 5.7. Exports of Petroleum, by Country, 1953–69 (Million Barrels)

Year[a]	Kuwait[b]	Saudi Arabia[b]	Iran[c]	Iraq	Libya	Venezuela
1953	312.7	302.5	—	204.6	—	606.5
1954	348.6	346.7	9.8	221.7	—	654.9
1955	400.3	350.4	110.7	239.6	—	738.6
1960	608.0	473.8	356.2	339.0	—	982.7
1961	624.4	530.2	397.5	347.2	5.1	1,009.0
1963	749.7	638.2	499.8	402.8	167.2	1,121.9
1965	850.6	787.4	644.1	458.8	442.6	1,187.2
1967	904.4	1,002.1	893.7	424.0	627.1	1,226.9
1968	951.6	1,009.3	975.9	525.2	945.0	1,235.3
1969	1,005.7	1,157.7	1,158.5	529.2	1,131.5	1,244.9

SOURCE: *International Petroleum Industry/Middle East* (New York: International Petroleum Institute, 1968); and additional data provided by the International Petroleum Institute, Inc., New York.

[a] Figures for three of the four major producing operations—KOC, Aramco, and the Iranian Consortium—are on a crude and crude-equivalent basis (Kuwait beginning 1956). Iraq, QPC, ADMA, ADPC, and Libya are actual crude exports. Venezuelan figures are crude and product exports, excluding bunkers. Crude production is used for BAPco.
[b] Kuwait and Saudi Arabia include one-half Neutral Zone (Aminoil/Getty and Arabian Oil).
[c] Iran includes IPAC and SIRIP beginning in 1965.

Table 5.8. Payments[a] by Oil Companies to Governments, by Country, 1953–69 (Million Dollars)

Year	Kuwait[b]	Saudi Arabia[b]	Iran	Iraq	Libya	Venezuela
1953	191.9	227.9	—	163.4	—	486.0
1954	219.1	293.2	22.5	191.4	—	510.0
1955	307.0	287.8	92.5	206.5	—	596.0
1960	465.2	355.2	285.0	266.3	—	877.0
1961	64.3	400.2	289.9	265.5	3.2	938.0
1963	556.7	502.1	388.0	325.1	108.8	1,106.0
1965	671.1	655.2	514.1	374.9	371.0	1,125.0
1967	709.9	842.5	751.6	362.1	625.0	1,206.0
1968	765.6	965.5	853.5	476.2	952.0	1,253.0
1969	812.2	1,008.0	1,040.0	483.4	1,132.0	1,289.0

SOURCE: *International Petroleum Industry/Middle East*. Data on Iran are from table 3.1 and additional data provided by the International Petroleum Institute, Inc., New York.

[a] Payments are obligations for royalty and income tax for the year shown, and in the Middle East include relatively small amounts arising from export refining operations in Saudi Arabia, Kuwait, and Iran. Retroactive payments, where possible, are allocated to the year applicable.
[b] Kuwait and Saudi Arabia each include one-half estimated payments of Arabian Oil Company, and estimated payments for each country respectively of Aminoil/Getty.

however, is still in Iran, with about 633,000 barrels per day for crude, and about 134,000 barrels for cracking and reforming.[28]

On the whole, Iran is now one of the biggest exporters and leading crude-oil producing countries in the world, with significant proved reserves. Iran's revenues from oil, also currently among the highest in the world, are about $1,700 million, or about one-eighth of gross national product. The activities of the oil sector in Iran have reached new peaks, thereby providing some of the most essential ingredients of rapid growth. In order to enhance still further the contributions of oil to the national economy, several new agreements with revolutionary features have been concluded with American and European oil companies in recent years. Table 5.9 summarizes some of the major provisions of Iranian oil agreements since 1954.[29]

---

[28] In Europe, Italy has emerged as the leading country, with about 3 million barrels of crude refining capacity per day, followed by West Germany (2.54 million barrels per day), France (2.53 million barrels per day), and the United Kingdom (2.39 million barrels per day). *Oil and Gas Journal* (28 December 1970), p. 92.

[29] For an in-depth study of the financial implications of the recent agreements between Iran and foreign oil companies, see Mansour Forouzan, "Comparison of Iran's Revenues from Oil under the 50-50 and ERAP Agreements," *Tahqiqat-e Eqtesadi*, published quarterly by the Institute for Economic Research, University of Tehran (April-September 1969), pp. 1–39, in Persian.

Table 5.9. Summary of Iran's Oil Agreements in the Postnationalization Period

Company	Short Name	Companies Constituting Second Party	Effective Date	Area (sq. km.)	Cash Bonus (million dollars)	Production Bonus (million dollars)	Exploration Period (years)	Minimum Exploration Obligation (million dollars)	Duration: Original Years + Optional Years
I. *Consortium Agreement*[a]									
Iranian Oil Exploration and Producing Co.		B.P. (40%) Shell (14%) Esso (7%) Gulf (7%) Texaco (7%) Standard–Calif. (7%)	10/24/54	Initially 254,112 approx. Currently 189,212 approx.					25 + 15
Iranian Oil Refining Co.		Mobil (7%) C.F.P. (6%) Iricon (5%)[b]							
II. *Joint-Venture Agreements*									
Société Irano Italienne des Pétroles	SIRIP	Agip S.P.A.	8/27/57	22,700	—	—	12	22	25 + 15
Iran Pan American Oil Company	IPAC	Pan American Pet. Corp.	6/5/58	14,600	25	—	12	82	25 + 15
Dashtestan Offshore Petroleum Company	DOPCO	Bataafse Petroleum Maatschappij (Shell)	2/13/65	6,036	59.01	28	12	18	25 + 15

Iranian Offshore Petroleum Company	IROPCO	Tidewater Oil Co. Skelly Oil Co. Sunray Oil Co. Kerr-McGee Oil Industries Inc. Cities Service Co. Atlantic-Richfield Superior Oil Co.	2/13/65	2,250	40	—	12	16	25 + 15
Iranian Marine International Oil Co.	IMINOCO	Agip S.P.A. Phillips Petroleum Co. Oil and Natural Gas Commission (India)	2/13/65	7,960	34	10	12	48	25 + 15
Lavan Petroleum Co.	LAPCO	Atlantic Richfield Co. Murphy Oil Corp. Sun Oil Company Union Oil Company	2/13/65	8,000	25	6	12	15	25 + 15
Farsi Petroleum Co.	F.P.C.	Bureau de Recherches de Pétroles Société National des Pétroles D'Aquintaine Regie Autonome des Pétroles	2/13/65	5,759	27	2	12	22	25 + 15
Persian Gulf Petroleum Company	PEGUPCO	Deutsch Erdoel-Aktien-gesellschaft, Deutsche Schachtbau und Tiefbohrgesellschaft Gelsenkirohner Bergwerke A.G. Gewerkschaft Elweath Preussag A.G. Scholven Chemie, A.G. Wintershall A.G.	15/7/65	5,150	5	5	12	10	25 + 15

(continued)

141

Table 5.9. (*continued*)

(continued)

III. Contract Agreements[c]

ERAP	The French State Co.	1966	275,500[d]	—	—	6	13	25
AREPI	European Consortium[f]	1969	27,260[e]	—	—	8	10	25
CONOCO	Continental Oil Co.	1969	21,860[e]	10[g]	—	7	12	25

SOURCE: *Iran Oil Mirror 1966–68*, pamphlet published by the National Iranian Oil Company, and *International Petroleum Encyclopedia* (Tulsa: Petroleum Publishing Company, 1969).

a The two (Exploration and Producing, and Refining) companies, registered as Dutch corporations are, in turn, controlled by the Iranian Oil Participants, Ltd. (registered under British laws). The consortium members have also established the Iranian Oil Services, Ltd. (also registered in Britain) and several trading companies for marketing and other ancillary services.

b Consisting of (1) American Independent Oil Company, (2) Atlantic Richfield Company, (3) Continental Oil Company, (4) Getty Oil Company, (5) Signal Companies, and (6) Standard Oil Company (Ohio).

c Under the "contract" agreements the National Iranian Oil Company retains 100 percent ownership of petroleum at the wellhead and the foreign companies act only as contractors. The latter are reimbursed by the right to purchase 30 to 45 percent of the oil produced after setting aside 50 percent as "national reserve" for Iran.

d 254,000 sq. km. offshore.

e Offshore.

f Consisting of France's Ente de Recherches et D'Activités Pétrolières (ERAP) (32%); Italy's Ente Nazionale Idrocarburi (ENI) (28%); Spain's Hispanoil (20%); Belgium's Petrofina (15%); and Austria's Oestreichische Mineraloelverwaltung A.G. (OMV) (5%). Mention must also be made here of the agreement signed with the Canadian Company, Sapphire Petroleum, Limited in 1958, which was later cancelled because of unsatisfactory progress in operations. But some of the other concerns (especially SIRIP, IPAC, and LAPCO) have already struck oil in viable commercial quantities.

g Bonus payment is broken down as follows: $1 million on signature, $1 million each of the following four anniversaries, $1 million on striking commercial quantities, $2 million when output reaches 100,000 b/d, and $2 million when output reaches 150,000 b/d.

Part 3

6

A Suggested Model

This study has been essentially concerned with the aggregate demand-supply relationships under dualistic conditions characterized by a dynamic, foreign-oriented enclave superimposed upon a traditional, nonindustrial economy. We have used Iran for our case study. In conducting our inquiry into the Iranian experience, we first took up the forty-year period preceding the nationalization of oil, followed by the more recent developments during the postnationalization years.

Our analysis of the available data indicates that during the 1910–50 period, the oil sector remained virtually divorced from the rest of the Iranian economy—a conclusion which is somewhat inconsistent with the Hirschman doctrine of unbalanced growth, but which has often been voiced and empirically investigated by others. However, the post-1951 period has been found to be at odds with the preceding forty years. In recent years, not only have the oil revenues accruing to the government been largely funneled into productive investments, but the intersectoral flows of resources have substantially increased—a conclusion which is in agreement with the general Hirschman thesis, and contrary to the conclusions reached otherwise.

In searching for the causes of the significant variations observed in the two periods, we have found that the government's role in the development process and the policies pursued by the political leadership, as well as the private response to those policies, are of utmost importance. Given appropriate policies on the part of the government, we have reached the conclusion that a foreign-financed enclave may indeed contribute to the development of the domestic economy in two important ways: (1) by the payments made by foreign concessionaires to the government (indirect influences); and (2) by the low-cost resources made available to the rest of the economy and the demand made on other domestic goods and services (direct influences).

The role of the government is thus of paramount importance in our suggested generalized model of development under dualistic conditions. This model will incorporate and bring together some of the important features and relationships we have observed in this study of a dominant modern enclave (the "dynamic sector") surrounded by a generally slower-moving economy (the "static sector").[1] Several reasons can be advanced in support of our probe. First, there is a need for a fresh theoretical approach to economic dualism, because evidently "neither theories of economic growth for an advanced economy nor theories of development for a backward economy are directly applicable to the development of a dual economy."[2] Second, there seems to be "a growing recognition that

---

[1] The division of the economy into two sectors is, of course, an abstraction. In reality, as Boeke and Schatz have emphasized, there are many sectors and subsectors displaying different degrees of progressiveness and dynamism. Our abstraction, however, has the advantage of focusing attention on the central feature of dualism, which is the coexistence of a large traditional sector with an active and dynamic industrial sector. Moreover, under dynamic conditions, both sectors undergo technological changes as they "interact" during the growth process. It is the interaction of these two sectors which is of central importance to dualistic economies, from which attention may be diverted by taking account of the many sectors that are encountered under actual conditions. It should be further noted that the use of the terms "static" and "dynamic" is in no way intended to imply that one sector remains stationary while the other grows, but rather to delineate the two sectors of a dualistic economy in the discussion that follows. See J. H. Boeke, *Economics and Economic Policy of Dual Societies* (New York: Institute of Pacific Relations, 1953), and S. P. Schatz, "A Dual Economy Model of an Underdeveloped Country," *Social Research* (Winter 1956), pp. 419–32.

[2] Dale W. Jorgenson, "Testing Alternative Theories of the Development of a Dual Economy," in *The Theory and Design of Economic Development*, ed. Irma Adelman and Erik Thorbecke, p. 45 (Baltimore: Johns Hopkins Press, 1966).

analysis of the economic interactions among sectors is the most promising approach to the study of development."[3] Third, what theoretical work has been published on dualistic economies has been largely concerned with the interactions between a subsistence agricultural sector and a dynamic commercialized-industrial sector, with the result that attention has been focused on "how to shift the economy's center of gravity from agriculture to industry until agriculture becomes a mere appendage."[4] Thus, the properties of dualistic economies characterized by a modern "enclave" implanted by advanced economies in an otherwise underdeveloped economic setting still remain to be more independently explored.

The structural characteristics of dual economies like Iran are somewhat different from those customarily found in the existing literature. The basic problem in our dual economies is not one of shifting the economy's center of gravity from the static to the dynamic sector, but of disseminating the growth-stimulating effects of the dynamic sector throughout the traditional sector by transforming the dynamic center into an engine of growth. This may or may not involve a shift of the economy's center of gravity toward the dynamic sector, depending on the conditions outlined below. Under most practical conditions, the weight of the argument seems to fall in favor of shifting the gravitational center away from the dynamic sector rather than toward it, as emphasized by contemporary theories. Our dual-economy model thus differs from those developed by Fei, Ranis, Jorgenson, and Schatz in two important respects: (1) the direct contact between the two sectors in our model is embodied in the flow of resources from the dynamic to the static sector, whereas in their models the flow of resources is generally in the opposite direction with the ultimate objective of shifting the economy's center of gravity from the static to the dynamic sector; (2) in our model the indirect (fiscal) influences of the dynamic sector provide the most important link between the two sectors, whereas in other models this is not explicitly significant.

---

[3] Robert E. Baldwin, "Comment," in *Theory and Design of Economic Development*, p. 41.

[4] John C. H. Fei and Gustav Ranis, "Agrarianism, Dualism, and Economic Development," in *Theory and Design of Economic Development*, p. 23. See also Schatz, "Dual Economy Model," for a description of a dual-economy model emphasizing the interaction between a primitive agricultural sector and a dynamic commercialized sector.

The Model's Basic Relationships

We may now begin our analysis by noting that in a dual economy of the type we have discussed [5] the dynamic sector acts as the "leading sector" or "growth center" which contributes to the development of the entire economy. However, the manner in which the dynamic sector performs this function is generally very different from that of the traditional leading sectors of the Western growth models. The difference arises largely from the fact that in the Western economies the leading sector is essentially an indigenous sector closely interwoven with the other sectors of the economy, whereas in dualistic economies the leading sector is largely alien to the other sectors, as if it were superimposed on the domestic economy. It is for this reason that the leading sector under dualistic conditions does not induce the crablike responses postulated by the Hirschman thesis. It is also for this reason that the direct influences of the dynamic sector tend to be insignificant, at least in the short run.

The relative insignificance of the direct influences of the dynamic sector, as has been substantiated by empirical evidence here and elsewhere, means that the only connecting link between the dynamic and indigenous sectors is, for all practical purposes, the indirect (fiscal) influences that flow from the dynamic sector into the public treasury and on into the domestic economy. The interdependencies of the dynamic and indigenous sectors via direct influences are bound to be trivial, although they need not remain so forever. The magnitude of direct influences may, under certain circumstances, logically grow and even supersede that of indirect influences in the long run. This outcome, however, may, in most cases, be far-fetched and unachievable within a reasonably short time.

Our suggested model consists of nine basic relationships, concerning the production function in the dynamic sector, the surplus generated in that sector, the division of the surplus between domestic income and factor payments abroad, the rate of growth of the dynamic sector, the production function in the static sector, accretion of surplus in the static sector, the rate of growth in this sector, and, finally, the capacity of the static sector for absorbing not only the dynamic sector's surplus, but also the actual physical product of the latter. Symbolically:

---

[5] It may be noted that unless otherwise stated the term "dual economy" will refer to the coexistence of a large traditional sector and an advanced export-oriented sector established and nurtured by foreign concessionaires.

(1)
$$A(t) = e^{\int \beta dt} K'(t)$$

(Production function in the dynamic sector where $A(t) =$ output; $K'(t) =$ capital; $\beta =$ innovation intensity in dynamic sector)

(2)
$$T(t) = A(t) - C(t)$$

(Total surplus in the dynamic sector; $T(t) =$ surplus; $C(t) =$ cost)[6]

(3)
$$T(t) = T'(t) + T''(t)$$

(Allocation of surplus between the host government, $T'(t)$, and concessionaire, $T''(t)$)

(4)
$$\beta = f(T''(t))$$

(Intensity of innovation in dynamic sector—endogenously determined)

(5)
$$B(t) = F(g(t)L(t), h(t)K(t))$$

(Factor-augmenting production function in the static sector; $B(t) =$ output; g and $h =$ augmentation terms applicable to L and K—labor and capital)

(6)
$$J = a + bc^{-ut} \qquad a \geq 0, b \geq 0, u \geq 0$$

(Intensity of innovation assimilation in static sector—exogenously determined)

---

[6] It must be noted that here we are abstracting from the flow of resources (such as labor and capital) from the indigenous sectors to the dynamic sector on the grounds that the use of local labor in the dynamic sector in general tends to be trivial and the capital equipment required by the dynamic sector is beyond the capability of the domestic economy to produce. The flow of resources will, therefore, be mainly in the form of raw materials from the dynamic sector to the various sectors composing the domestic economy. For a similar abstraction see Erik Thorbecke and Apostolos Condos, "Macroeconomic Growth and Development Models of the Peruvian Economy," in *Theory and Design of Economic Development*, pp. 181–208. This abstraction also implies that the costs of production denoted by $C(t)$ consist of payments to foreign factors of production only. This assumption of zero domestic factor earnings is consistent with the high capital-intensive nature of the dynamic sector and the absence of a capital-goods producing sector in the domestic economy. Where this assumption does not hold (e.g., tourism), the link between the two sectors will be provided largely through C, in which case the Fei-Ranis-Jorgenson-Schatz dual-economy theories, modified to account for a dynamic service sector instead of an industrial one, will probably apply.

(7) $$T_s(t) = lB(t)$$

(Surplus in static sector)

(8) $$\eta K(t) = (T_s(t) + T'(t))/K(t)$$

(Rate of growth of capital in static sector)

(9) $$\eta A(t) = f(\eta K(t), J, U(t))$$
$$\partial \eta A(t)/\partial \eta K(t) \geq 0$$
$$\partial \eta A(t)/\partial J \geq 0$$
$$\partial \eta A(t)/\partial U(t) \geq 0$$

(Absorption of output of dynamic sector in static sector; U incorporates all independent variables not contained in $\eta K(t)$ and J)

The basic production conditions of the dynamic and static sectors of a dual economy are given by equations (1) and (5) respectively. For the former we have postulated a production function with no labor as an independent variable, and for the latter we have assumed a factor-augmenting function.[7] The innovation intensity for the dynamic sector (equation [4]), is—somewhat optimistically—assumed to be endogenously determined as a function of the surplus after allowing for the share of the host government (i.e., indirect influences). The intensity of innovation assimilation of the static sector, subject to certain qualifications, is assumed to be exogenously determined, which is consistent with the Veblenian "latecomer-nation" hypothesis.[8] According to this hypothesis, the static sector of a dualistic economy arriving on the world technological scene as a latecomer,

---

[7] We have used this form of production function deliberately to allow for varying shares of labor and capital in total output as factor proportions change. The commonly used Cobb-Douglas production function (see Fei and Ranis, "Agrarianism, Dualism, and Economic Development," p. 29) assumes unitary elasticity of substitution and Hicks-neutrality of innovation with constant proportionate product-raising effect per period and constant relative shares. Such a function does not seem to be consistent with dynamic development where innovating activity need not be Hicks-neutral. See William Fellner, "Measures of Technological Progress in the Light of Recent Growth Theories," *American Economic Review* (December 1967), pp. 1073–98.

[8] Thorstein Veblen, "The Opportunity of Japan," *Essays in Our Changing Order*, ed. Leon Ardzrooni (New York: Viking Press, 1954), pp. 248–66, reprinted from *The Journal of Race Development* (July 1915).

and eager to borrow technology from abroad, will behave as postulated in (6). This means that factor J will initially start at a given level and then monotonically decrease to a stationary level a, as the advantages of importing technology gradually decline.

Equation (8) describes the rate of capital formation in the static sector, which explicitly takes into account the indirect influences flowing from the dynamic sector. As can be readily surmised, the rate of capital accumulation in this sector is directly related not only to the magnitude of surplus transferred from the dynamic sector, but also to two other variables tacitly implied in our equation: (1) the quality of national development planning and implementation; and (2) the resources potentiality of the static sector for growth independent of the dynamic sector.[9] This is so because the stock of the real physical capital can grow only if the foregoing two conditions are also present. Thus, countries like Iran (as with Iraq, Venezuela, Indonesia, and perhaps Algeria among the oil-producing underdeveloped countries), that have undeveloped resources other than oil can expect a faster rate of overall growth than, say, countries like Abu Dhabi, Kuwait, and Saudi Arabia, that as yet show no significant resources other than oil. The surpluses generated in the dynamic sector of the latter countries tend to be largely spent on imports of consumer goods or accumulated as foreign exchange reserves in foreign banks, due to the lack of viable domestic investment opportunities.

Equation (9) is more important for our purposes because it incorporates and brings together in a functional form the direct and indirect influences of our dual-economy model. The flow of resources from the dynamic sector to the static sector in the form of raw or processed materials is assumed to depend mainly on the absorptive capacity of the static sector (as represented by its rate of capital formation) and the intensity of innovation assimilation in that sector. Thus, the magnitude of direct influences may increase as the stock of capital in the static sector increases and as the rate of innovation assimilation is intensified.

Given our assumptions regarding the capital-using, labor-saving

---

[9] Here we are implicitly assuming that all the surplus transferred to the static sector (i.e., indirect influences) is added to the stock of capital. This assumption is only for the sake of analytical convenience and may be modified to take into account the portion of the surplus that is consumed without affecting our analysis.

nature of the dynamic sector, and the incapability of the domestic economy to supply the input requirements of the dynamic sector (except, of course, the basic raw material, which is not relevant to our model, and a relatively small amount of labor, which is immaterial for our purposes), the intersectoral flow of resources nurtured by changes in demand-supply relationships will be essentially a one-way flow from the dynamic to the static sector.[10] If the absorptive capacity of the static sector is initially large or if it grows over time (that is, if $\partial \eta A(t)/\partial \eta K(t) > 0$, $\partial \eta A(t)/\partial J > 0$, and $\partial \eta A(t)/\partial U(t) > 0$), then the flow of resources in the form of raw and intermediary products from the dynamic sector to the static sector will rise. On the other hand, if the output of the dynamic sector is initially unabsorbable in the static sector and resists assimilation over time (that is, if $\partial \eta A(t)/\partial \eta K(t) = 0$, $\partial \eta A(t)/\partial J = 0$, and $\partial \eta A(t)/\partial U(t) = 0$), then the interaction of the two sectors via direct influences will be almost nil; and the dynamic sector will consequently remain relatively isolated indefinitely.

The magnitude of direct influences, by implication, depends largely on the versatility of the output of the dynamic sector itself (or, more precisely, the propensity of the static sector to use the output of the dynamic sector). The more widely usable the output (i.e., the greater the static sector's propensity to absorb), the greater will be the extent of direct influences. Skilled labor, for example, is most versatile; uranium is least; petroleum stands somewhere in between. Thus, if the product of the foreign-related dynamic sector should be skilled immigrants—as has, so far, been the case with Israel—the rest of the economy could absorb the dynamic sector's output more easily and in large doses and the impact of the direct influences will be extensive. If, on the other hand, the output should be of such a nature as to have very little (primary or secondary) use in the domestic economy—like tin, copper, cocoa in Bolivia, Zambia, and Ghana—the situation would be vastly different. In such relatively small, not easily diversifiable, countries producing an export-oriented material with a small degree of absorbability into the home economy, the direct impact of the growth sector on the rest of the economy would be limited, and whatever influences there are would be *indirect*. In a tiny, barren country with no other

---

[10] For an empirical verification of this conclusion and the underlying assumptions, see our discussion concerning direct and indirect influences of the oil sector in Iran, chaps. 2 and 3.

resources to develop but oil, petroleum, too, would fall into the same category; that is, with almost no influences beyond increasing the treasury's foreign exchange holdings.[11]

The Model's Implications

Our suggested model presents several important implications for dualistic growth. First, as was indicated before, the assumption that direct influences are a one-way flow from the dynamic sector to the static sector is contrary to the assumption usually made in discussing intersectoral flows under dualistic conditions not characterized by a foreign-oriented enclave. Under the latter conditions, the dynamic sector grows by drawing upon the resources freed from the static sector, until the former emerges as the dominant sector and the latter becomes a mere appendage. In dualistic situations of the type we have discussed in this study, the major part of the flow of resources is not from the static sector to the dynamic sector, but the reverse. Whether the continuation of this process will eventually dwarf the dynamic sector into a mere *appendage* of the static sector—an outcome which is contrary to that envisaged under conventional dualistic conditions—depends, as was argued before, upon the absorptive capacity of the static sector and the rate of increase of that capacity over time.

To put this differently, if the expansion of the static sector is partly nurtured by the output of the dynamic sector, the speed with which the output of the latter sector will be absorbed by the former sector will depend on the intensity and extensity of existing and emerging linkages. Should such linkages become sufficiently significant over time, the dynamic sector may ultimately become largely integrated into the domestic economy and consequently lose its dominant position in favor of the static sector. This conclusion does not, of course, imply that the dynamic sector will cease to be export-

---

[11] The case where the absorption of the output of the dynamic sector is a positive function of the variables set forth in equation (9) is exemplified by the recent experience of Iran with oil. The flow of oil and related products to the various sectors of the Iranian economy has sharply increased in the past decade and is expected to continue to rise in the future. This has been accompanied by capital accumulation and assimilation of innovation in the domestic economy, as reflected in the emergence of productivity points and the resulting shifts in the importance of various sectors in national output.

oriented and become only a supply unit of the growing but originally static sector. What it does imply is that the behavior of the dynamic sector may eventually tend to conform to the more traditional export sectors where a substantial market exists on the home front but is supplemented and sometimes substantially augmented by foreign demand.

Second, we have concluded that if the output of the dynamic sector and the nature of the static sector are such that the former is completely alien to the latter and continues to remain unabsorbable over time, then the dynamic sector will always be independent of the static sector and the only connecting link between the two will be the direct transfer of financial resources (indirect influences) from the dynamic sector to the static sector via the public treasury. Such an outcome, however, except for some tiny oil-producing countries and territories in the Persian Gulf, is rather improbable in the long run. That is, even if the two sectors under a dualistic situation of the type we have discussed do not intermingle initially, or remain divorced for some time, it would be rather unlikely for them to remain completely independent of and indifferent to each other in the long run, unless the static sector (as in the Persian Gulf sheikdoms) remains truly stagnant with no change in resource endowments, including entrepreneurial ability, or an indifferent government. Where the output of the dynamic sector consists of resources with a high degree of absorbability, our model indicates an increasing flow of resources from the dynamic sector to the static sector, with the result that the growth of the latter will be aided and expedited by the former.[12]

Third, the shift of the economy's gravitational center away from the dynamic sector in favor of the static sector not only is likely to be brought about by the flow of resources from the former into the latter, but is apt to be strongly influenced by the indirect (fiscal) influences of the dynamic sector and the set of policies pertaining thereto. Of crucial importance here is the contractual and legal

---

[12] This conclusion follows readily from a simplified version of our model. Let R represent a given rate of growth of output in the static sector (B), E the output-elasticity of demand for the output of dynamic sector, and m the marginal propensity to consume the output of dynamic sector; it can be shown that the increase in demand for the output of the latter sector is $m(dB/dt)$ and the rate of increase of demand is ER. See Harry G. Johnson, *International Trade and Economic Growth* (Cambridge: Harvard University Press, 1961), p. 66.

framework within which foreign concessionaires are supposed to operate. The nature of the concession—particularly the privileges given to foreign interests (e.g., duty-free imports) by which they can keep the dynamic sector isolated from the rest of the economy—can make all the difference.[13] This is another aspect of what we referred to earlier as the institutional constraint of our model. Equally important is the degree to which a government, bent on economic development, is willing to channel all or most of its receipts from the dynamic sector into development projects in which the country enjoys a comparative advantage. The latter, as has been pointed out before, is likely to be affected by the low-cost source of supply made available by the dynamic raw-materials sector. The immediate and potential impact of this sector upon the economy's cost and production patterns will depend, among other things, on the relative importance of the raw materials in the overall costs of existing and emerging industries. If the availability of low-cost raw materials is such that it alters the production costs and hence the comparative advantage of the country toward the production of those goods that use the raw materials relatively intensively, then the channeling of funds by governmental action is likely to result in an increased flow of resources from the dynamic to the static sector.[14]

We do not mean to give the impression here that the direct resource flow or assimilation of the dynamic sector with the rest of the economy is of paramount priority in the development strategy. In fact, where large domestic or foreign markets are needed in order to develop

---

[13] In the case of Iran, for example, under the 1954 agreement, Iran was allowed to sell its 12.5 percent share of crude oil to the consortium at the posted price in the Persian Gulf (which is generally higher than actual prices received by the consortium for the sale of oil). But if NIOC wished to take the crude and market it independently, it had to pay a fee of 2.3 percent to the operating companies. Thus, during the period of financial difficulties, the Iranian government chose to receive cash instead of crude—giving up the chance of finding new independent markets abroad. Since the consortium is also not obligated to refine all of the Iranian crude in Iran, it chooses to concentrate on the export of crude petroleum.

[14] Our examination of the data on Iran suggests that this has been, especially in recent years, taking place in the Iranian economy. The establishment and vigorous expansion of petrochemicals and energy-intensive industries in Iran, as has been indicated previously, are probably indicative of the impact of the oil industry on the pattern of costs and the economy's comparative advantage.

industries that can use the products of the dynamic sector, the priority may not be very high and there may be definite limits to the intersectoral flows between the two sectors; opportunities in other fields may be more attractive.

Fourth, as can be concluded from our arguments, much of what we have said concerning the intersectoral relationships in a dualistic economy revolves around the role of the government. Unlike a Schumpeterian model where the generating force of development is provided by the entrepreneur, our model places the government in the vanguard of development. This is because the development process in most developing economies is to a much larger extent imbued with political, social, cultural, and nationalistic considerations.[15] That the role of the entrepreneur may be secondary in the development process of many developing countries is suggested by the experience of Iran. One of the reasons for the Iranian economy's rather slow growth during the first forty years after the establishment of the oil industry was probably the country's insufficient endowment with private entrepreneurial talent. Yet one of the reasons was also probably the relatively unsophisticated nature of the national development program. Although both the entrepreneurial endowment of the country and the role of the government in the economy have undergone substantial changes during the postnationalization years, there is no doubt that the government has been the main instigator and a major supporter of the development effort.

The substitution of government for entrepreneurial initiative has, incidentally, its own implications for development theory. For one thing, it is no longer necessary to postulate innovation as the characteristic feature of the development process. Instead, the government may in effect take the lead in importing technology from abroad and disseminating it throughout the economy. This process of development by borrowing, absorbing, and assimilating technology from abroad is, as Wallich has aptly pointed out, very dissimilar to the traditional innovation processes. For another thing, if the entrepreneur is not assumed to be the propelling force of development, neither can his profit be considered among the important variables in the development process. Instead, the preference function of the government, derived from the preference schedule of the people for higher living

---

[15] See Jahangir Amuzegar, "Nationalism versus Economic Growth," *Foreign Affairs* (July 1966), pp. 651–61.

standards, would become the relevant variable.[16] The obvious corollary of this argument is a shift in the development orientation itself from production and supply (which is characteristic of the Schumpeterian model) to consumption and demand. This, in Wallich's view, is primarily due to the conscious efforts in many developing economies to substitute (1) the government for the traditional entrepreneur, (2) foreign-borrowed technology for the original process of innovation, and (3) higher living standards for the entrepreneurial profit among the primary development goals. All this, in turn, tends to predispose the developing economy toward industrialization, because higher living standards are heavily biased toward an industrial and urban way of life, and the assimilation of foreign technology is perhaps easier and more productive in mining and manufacturing (or even mechanized agriculture) than in other fields of development, such as education or public administration.[17]

It should be noted at this point that the essentiality of public initiative and finance in the design and direction of development under dualistic conditions is also an inherent source of frustration for the normal growth of the economy. That is to say, those who are in a position to make the economy can break it too. The vital dependence of high domestic growth rates on incomes from the dynamic industry places the behavior of this industry (and the planners' preferences based on its receipts) at the core of all developmental efforts. But, since the revenues from an export-oriented industry depend on world market demand and supply and are largely beyond the planners' calculation and control, the possibilities of miscalculation and inadequate control usually loom large. The planners' exaggerated ambitions for growth, or their overzealousness in outsmarting themselves, may push the economy beyond the brink of safety or normality.

These possibilities are particularly abundant in countries without comprehensive state planning and operation where the private

---

[16] In our own model the substitution of government for the entrepreneur means that equations (6), (8), and to a large extent (9) are directly influenced by governmental action. This, in turn, means that the pace and tempo of innovation assimilation, capital formation, and consequently the flow of resources from the dynamic to the static sector are by and large set by the government. Given the predisposition of the economy toward industrialization as reflected in the preference function of the government, the speed and content of industrial advancement will also be determined by the state.

[17] Wallich, "Theory of Derived Development," pp. 190–204.

sector still plays a crucial role in the economy's overall performance. In such countries, like Iran, those who help push the applecart ought to be careful not to upset it. Since each rial of projected oil revenues received or not received, spent or not spent, rightly invested or malinvested presents far-reaching consequences for the entire economy (over much of which the state has little or no direct control) a particular kind of wisdom and foresight on the part of the planners is absolutely essential. In centralized economies with comprehensive planning, the state is usually in a position to revise its projections downward or upward; and act accordingly, with far greater assurance of success and with a minimum of unforeseen disruptions.

Some Further Elaborations

As was shown for Iran, the problems ordinarily encountered as a result of industrialization under nondualistic conditions (particularly insufficiency of savings and a secular tendency toward inflation) need not become critical factors under dualistic conditions. As our model indicates, the indirect (fiscal) influences directly affect the rate of capital formation. To the extent that these receipts do not involve the usual sort of belt-tightening by consumers, they may be considered a perfect source of investable funds. The revenues thus accruing to the government supplement internal savings in an important sense and thereby alleviate the ubiquitous problem of savings scarcity. Moreover, since these revenues are in foreign exchange and readily acceptable everywhere, the often painful and sometimes impossible task of converting domestic savings into an internationally acceptable means of payment is avoided.[18] As has been observed earlier in this study, this has further implications for aggregate consumption and investment as well as for balance-of-payments considerations and monetary and fiscal policies. The relationships between foreign exchange receipts of the government from the dynamic sector and aggregate savings and investment in the indigenous sector are shown in the following set of equations comprising total resources available (R), national income generated in the static sector (Y), consumption (C), investment (I), savings invested

---

[18] Jahangir Amuzegar, "Atypical Backwardness and Investment Criteria," *Economia Internazionale* (August 1960), p. 11.

domestically (S), imports (M), exports (X) (excluding the exports of the dynamic sector), indirect influences (T'), and a foreign exchange gap (D).[19] We have:

(10) $\quad\quad R = C + I + X \quad\quad$ (aggregate demand for resources)

(11) $\quad\quad R = Y + M \quad\quad$ (aggregate supply of resources)

(12) $\quad\quad M = X + T' + D \quad\quad$ (foreign exchange position)

(13) $\quad\quad Y + T' = C + S \quad\quad$ (disposition of income)

From these equations, we have:

(14) $\quad\quad\quad\quad I = S + D \quad$ or $\quad D = I - S$

and

(15) $$D = M - (X + T')$$

As can be seen from these equations, the most proximate contribution to growth of periodic receipts by the government from the dynamic sector is its direct impact on financing imports. This has an undeniably material effect in narrowing the exchange gap and, finally, in alleviating the burden of otherwise heavy debts that would be necessary to meet the rapidly mounting requirements of the domestic economy. This in effect provides badly needed relief from the perennial problem of having to produce enough exportables to pay for imports. This obviously affords the country a higher level of real consumption than would be obtainable otherwise. Moreover, it has the additional long-run advantage of rapidly increasing the productive capacity of the economy, thereby exerting a dampening influence on inflationary pressures stemming from the desire of individuals for higher living standards.[20]

An interesting conclusion emerges from equations (14) and (15), which imply equality between the savings gap $(I - S)$ and the

---

[19] The equations presented here are a modified form of the set of equations in Douglas S. Paauw and Forrest E. Cookson, *Planning Capital Inflows for Southeast Asia* (Washington: National Planning Association, 1966).

[20] As is clear from the foregoing analyses, we are assuming that the foreign exchange made available by the dynamic sector is used to import capital goods, not consumption goods. This is consistent with our previous assumption that the indirect influence of the dynamic sector directly augments the economy's stock of capital and is further borne out by the current structure of Iranian imports in which capital and intermediary goods together account for almost 90 percent of the total value of imports.

foreign exchange gap $(M - (X + T'))$. This results from the requirement of balanced growth, which would seem to imply no divergence
between the two gaps. It should be noted here that the behavior
of the foregoing variables has been based on certain implicit behav-
ioral relationships which may now be made explicit as follows:

$$M = mY$$
$$X = X(0)e^{\eta X(t)}$$
$$T' = T'(0)e^{\eta T'(t)}$$
$$Y = Y(0)e^{\eta Y(t)}$$

Hence,

$$D = mY(0)e^{\eta Y(t)} - (X(0)e^{\eta X(t)} + T'(0)e^{\eta T'(t)})$$

where $X(0)$ = level of exports at time $t = 0$,
$\eta X(t)$ = constant growth rate of exports,
m = ratio of imports to output,
$T'(0)$ = level of indirect influences at time $t = 0$,
$\eta T'(t)$ = constant growth rate of indirect influences,
$Y(0)$ = output at time $t = 0$, and
$\eta Y(t)$ = constant growth rate of output.

Here, we are assuming that the exports of the indigenous sector,
and the time path of the indirect influences, are both given exogenously; and that imports are related to the level of output. Both of
these assumptions, we think, are reasonable. First, the level of exports
of the indigenous sector is generally determined by prevailing conditions in the world market and can, therefore, be treated as a
given datum.[21] Similarly, assuming that the level and growth rate
of indirect influences are in general contractually established and
subject to external demand, it is also reasonable to assume that
they are exogenously determined. Second, the assumption that
imports are related to output is also justifiable on the grounds
that as development proceeds, a diversified pattern of goods will
be required to accommodate growth.

---

[21] This assumption has been found to be largely true of underdeveloped
countries in general. In a study of the export experience of underdeveloped
countries, Barend A. de Vries concludes that "there was no association
between variations in domestic price inflation and the performance of major
exports. Changes in exports . . . appeared to be associated with countries'
shares in the markets of major commodities . . . countries with small market
positions tended to fare better than those with large shares." See Barend
A. de Vries, *The Export Experience of Developing Countries*, World Bank
Staff Occasional Papers no. 3 (Baltimore: Johns Hopkins Press, 1967), p. 60.

Now, given the independent variables underlying the foregoing three functions, a divergence between imports and exports can easily emerge, since there is no a priori reason why the various functions should move along a zero-gap time path. In terms of our accounting equations, this may mean a continuous rise in M with possibly a discontinuous or slower rise in $T' + X$, thereby widening the gap. Although it may be difficult to exert any control over the trajectory of X in the short run, it may prove somewhat less difficult to adjust the magnitude of indirect influences through appropriate changes in the terms of the foreign concessions. This makes the dynamic sector more readily accessible as a source of maintaining balance between the two gaps. That is, given the difficulty of controlling the foreign exchange receipts from traditional exports, the governments of dualistic economies are likely to turn to the raw-materials-producing dynamic sector for the maintenance of the required balance between savings and investment. This would, in turn, involve periodic negotiations with foreign concessionaires to effect appropriate changes in the terms of existing contracts and liberalizing the terms of future contracts.

Policy Considerations

Although our intention in this study has not been to suggest any specific development policy, some policy considerations inevitably follow from ours. First, the fact that foreign investments in the export-oriented sectors of most underdeveloped countries have not, as a general rule, contributed much to general economic growth via direct influences has perhaps been due partly to the conditions inherent in the underdeveloped economies themselves, and partly also to the "exploitative" policies pursued by foreign concessionaires. That is, policies pursued deliberately or otherwise by foreign interests to insulate the dynamic sector from the rest of the economy and thus preclude integration of the raw-materials sector into the indigenous economy. In fact, one can think of many methods by which such an integration could have been expedited by foreign investors.[22] Yet

---

[22] Such methods could perhaps include encouraging and training a local pool of contractors, participating in the development of other industries with strong linkages to the minerals sector, giving preferential treatment to domestic materials and products, encouraging and supporting the development of local capital markets, and other similar measures that might have given rise to a new dynamic—but indigenous—sector in the economy.

the important point is that the absence of direct influences may, at least initially, be largely independent of whether the development of mineral resources is carried out by foreign concerns or by local entrepreneurs (including the host government). Indeed one can think of many instances where native exploiters did more damage to the economy than foreign intruders.

Second, the fiscal influences seem to be crucial in the early phases of development, because the magnitude of such influences directly affects the rate and the speed with which the government can bring about the desired integration of the minerals sector with the rest of the economy. Fiscal influences have the important effect of substituting for other sources of revenue that the government would otherwise have to rely on to finance its development programs. The fiscal influences not only greatly relieve the government from employing the complex and tortuous means of raising funds by taxation, inflation, and so on, they have the unique effect of permitting the private demand (consumption and investment) to grow faster than it otherwise could. This means that consumers will be able to enjoy higher standards of living while maintaining relatively high rates of investment. Therefore, although foreign concessionaires may have been unable, at least initially, to alter the magnitude of direct influences, they could have had a direct and important role in determining the magnitude of fiscal influences and hence in accelerating or decelerating the economic growth of the host countries and the eventual integration of the relevant sectors into an interdependent whole.

Third, it follows from our analysis here that the strategy of the government must be "balanced" in the sense that the government should attempt to help diversify the economy and make its other industries expand and catch up with the fast-growing minerals sector. That is, industry, other minerals, agriculture, and services should all be emphasized in order to (1) increase the degree of integration between the various sectors, (2) supply food for the growing population, and (3) provide services for consumption by the individuals in the expanding industrial sector. In order to achieve such a diversification, and to reduce substantial dependence on the one dynamic sector, the investment criteria generally suggested for the allocation of resources in the underdeveloped countries may have to be revised, or at least supplemented with other criteria. One such criterion must, of course, be the conservation of the vital natural resource, consistent with its most profitable utilization. The linkage of new investment plans with the dynamic, foreign-financed sector must also be taken into account.

Fourth, it follows from the foregoing arguments that the long-run planning efforts of the government must be focused on creating those conditions that would help the indigenous economy to respond to the numerous supply-demand imbalances generated by the dynamic sector. However, given the sophisticated and highly capital-intensive nature of the dynamic sector (e.g., oil industry), it would be unrealistic, at least at the earlier stages of development, to gear the economy toward responding to the demand imbalances created only by that sector. The development strategy should, therefore, be aimed at taking maximum advantage of supply-induced imbalance (i.e., low-cost materials). This means encouraging and expanding in particular those industries that use the low-cost materials intensively (e.g., petrochemicals or energy-based industries like aluminum or other metal smelting and refining in the case of oil). Development efforts should, in other words, be directed first to expanding those industries whose efficiency and competitiveness are influenced by low-cost materials flowing from the dynamic sector, and second to creating new industrial growth centers that will eventually lead to sustained growth.

The successful implementation of this strategy is no mean feat. This phase of development planning is enormously difficult and complex, because it implies a thorough investigation of the economy's potentials in order to determine the areas of comparative advantage. Yet it is imperative that such areas be clearly determined so that the benefits of the fast-growing sectors may be somehow captured by other sectors, and so that the necessary momentum may be created for continued self-sustaining growth.

Basic Data on the Economy of Iran

Physical Characteristics

Area:
628,000 square miles distributed as follows:

Under cultivation (including fallow)	11.5%
Pastures	6.1
Forests	11.5
Cultivatable	18.8
Uncultivatable	52.1

Minerals:
Petroleum reserves: 70 billion barrels
Natural gas reserves: 214 trillion cubic feet
Copper: latest estimate of reserves in the Kerman region; 400 million tons
Other: zinc, lead, chromite, iron ore, manganese, nickel, cobalt, tungsten, gold, and silver

Marine:
Over 80 varieties of fish in the Caspian, including the caviar-bearing sturgeon
200 species in the Persian Gulf

Forests:
72,200 square miles, mainly oak, ash, elm, beech, poplar pine, box cypress, maple, walnut, and honey locust

Manpower

Population:	29.7 million
Birth rate:	Approximately 50 per 1,000
Death rate:	Approximately 20 per 1,000
Natural rate of growth:	2.86 percent
Density:	47.29 per square mile
Median age:	16.9 years
Mean age:	22.2 years
Average life expectancy:	46 years
Literacy:	35 percent
Per capita income:	$300 per year

Estimated Gross National Product (at factor cost), 1970 (1959 Prices)

GNP (billion rials)	*590.0*	*100.0%*
Agriculture	(120.0)	(20.3)
Oil	(116.9)	(19.8)
Industry and mines	(71.2)	(12.1)
Construction	(24.5)	(4.2)
Water and electricity	(16.0)	(2.7)
Transport and communications	(34.8)	(5.9)
Banking and insurance	(24.5)	(4.1)
Wholesale and retail trade	(57.0)	(9.7)
Ownership of dwellings	(33.0)	(5.6)
Other	(92.1)	(15.6)

Foreign Trade (million dollars), Annual Averages: 1962–70

Oil exports (posted prices)	*1,415.0*	
Nonoil exports	*188.6*	*100.0%*
Carpets	(48.0)	(25.4)
Cotton	(41.0)	(21.7)
Fruits	(27.1)	(14.4)
Other	(72.5)	(38.5)

Imports	*1,080.0*	*100.0%*
Machinery	(236.0)	(21.8)
Iron and steel	(180.0)	(16.7)
Chemicals	(92.9)	(8.6)
Electrical appliances	(84.5)	(7.8)
Other	(486.6)	(45.1)

Government Finance (billion rials), Annual Averages: 1963–70

Current revenue	111.2
Oil revenue	(56.1)
Nonoil revenue	(55.1)
Current expenditure	89.4
Capital expenditure	52.8
Overall deficit	31.0

Unit of currency	Rial (Rl.) = 100 dinars
Exchange rates	Rls. 76.5 = U.S. $1
	Rls. 183.3 = £1

Index

169